Geology of Petroleum

Edited by Heinz Beckmann

Vol. 2

Geological Prospecting of Petroleum

by Heinz Beckmann

A Halsted Press Book

John Wiley & Sons, New York — Toronto

Author:

Prof. Dr. Heinz Beckmann
Lehrstuhl für Erdölgeologie
der Technischen Universität Clausthal-Zellerfeld

Translation: *Donald McLeod*

Library of Congress Cataloging in Publication Data

Beckmann, Heinz, 1918—
    Geological prospecting of petroleum.

    (Geology of petroleum ; v. 2)
    Bibliography: p. VIII, 183
    Includes index.
    1. Petroleum--Geology. 2. Prospecting. 3. Oil well
logging. I. Title. II. Series.
TN870.5.G42 . vol. 2 553'.28'08 [622'.18'282] 76—156
ISBN 0—470—15209—5

Distributed in the USA, Canada and Latin America by Halsted Press, a Division
of John Wiley & Sons, Inc., New York.

© Ferdinand Enke Verlag, Stuttgart 1976

Printed in Germany by Printing House Dörr (Adam Götz, proprietor), Ludwigsburg

For Ferdinand Enke Verlag ISBN 3 432 02198 4

Preface

The science of petroleum geology is only about fifty years old, but the amount of literature on this subject has already become larger than that of some of the much older classical sciences. A young geologist who wants to learn something about oil well drilling will find it difficult to know where to begin reading and learning. Of course there are several adequate text books on wellsite petroleum geology, but most of them are too voluminous and too expensive for out door work. This booklet has been written to fill the gap between such big manuals and the short comments on practical petroleum geology found in the usual geological text books. The decision which theories and methods to mention here, and which to omit without detriment to the reader is a very difficult one, of course, and subject to individual preferences and practical experience in only some of the regions of the oil producing world.

The selection made in this booklet is based on about ten years of practical well-site geology and about fifteen years of teaching petroleum geology. The aim of this paper is mainly to show a young geologist what is going on at the drilling rig and what he is expected to do. Well-site geology means teamwork with the geophysicist and with the driller, later on perhaps with the production engineers and other specialists. So he has to understand their language, he has to know their methods and their limitations. On the other hand the paper will be useful for the team partners, too, giving them information on practical well-site petroleum geology and the limitations of these methods.

In general the paper has been written mainly for the practical geologist and has left forming theories on the origin of petroleum and similar problems to other authors.

Contents

1.	*History of petroleum and petroleum geology*	1
1.1	The use of petroleum during the ages	1
1.1.1	Petroleum and energy	3
1.1.2	Petroleum and economy	3
1.1.3	Petroleum and illumination	4
1.1.4	Petroleum and war	5
1.1.5	Petroleum and pollution	6
1.1.6	Petroleum and medicine	7
1.1.7	Petroleum and religion	9
1.1.8	Petroleum and food	11
1.2	Petroleum and the future of this kind of energy	12
1.3	History of petroleum exploration and of petroleum geology	12

2.	*Origin and accumulation of petroleum and natural gases*	15
2.1	Theories about the origin of petroleum hydrocarbons	15
2.2	The migration of petroleum hydrocarbons	18
2.3	Some words about the chemistry of petroleum hydrocarbons	20
2.4	The pay and its properties	25
2.5	The structure and contents of oil- and gas-bearing reservoirs	32
2.5.1	What does the pore space contain in reality?	
2.5.2	The technical properties and the commercial value of natural gases and crude oil	
2.5.3	Primer structural types of reservoirs	39
2.5.4	The most common types of reservoir in geosynclines and graben structures	50
2.5.5	Types of reservoir caused by salt domes	50
2.5.6	Reservoir caused by reefs	52
2.5.7	Reservoirs caused by buried hills	53

3.	*The structure of an average oil company*	55
3.1	Geological laboratories and specialists	57

4.	*The principal laws of petroleum exploration*	59
4.1	Some words about methods of petroleum exploration	60
4.1.1	The first steps of petroleum exploration	60
4.1.2	Big scale exploration by airborne magnetics	61
4.1.3	Surface gas logging	62
4.1.4	Big scale exploration by gravimetric surveys	63
4.1.5	Big scale exploration by refraction seismics	63
4.1.6	Small scale exploration by geoelectrics	64
4.1.7	Detail petroleum exploration by reflexion seismics	64
4.1.8	How to outline a saltdome in the underground	66
4.1.9	Mapping the underground by structural drillings	67

5.	*The practice of deep drilling*	68
5.1	What a geologist should know about drilling technique	70
5.1.1	The drilling rig and its most important parts	72
5.1.2	What a geologist should know about drilling fluids	90
5.1.3	Blow out and lost circulation, a mud engineering problem with some geological points of view	94

6.   *Mud gas logging*   97
6.1   Cold wire and hot wire detectors   97
6.2   Flame ionisation detectors 100
6.3   Surface active transistors used for gas logging 100
6.4   Pellistor detectors 101
6.5   Sulphide detectors 101
6.6   Infra-red detectors 101
6.7   Chromatographic mud gas analysis 105
6.8   Mud gas logging and blow out prevention 106
6.9   Logging the density of the drilling fluid 107
6.10   Gas tails and false gas alarm 108
6.11   How to analyse the gas contents of cuttings 109

7.   *Cuttings and how to analyse them stratigraphically* 112
7.1   Petrographic analysis of cuttings 112
7.2   Biochronological analysis of cuttings 112
7.3   How to extract microfossils from cuttings 118
7.4   Acetate films made from cuttings of hard formation 120
7.5   Insoluble residues made from cuttings 121
7.6   Dating cuttings by nannofossils 121

8.   *Cores and core analysis* 123
8.1   How to place a core properly 123
8.2   Direct methods of coring a pay 124
8.3   How to drill a core 125
8.4   Side wall coring on the logging cable 130
8.5   How to describe and to analyse a core 131
8.6   Testing the properties of sandstone cores 133
8.7   The origin of porosity and permeability 137

9.   *A short course on bore hole logging* 140
9.1   Logging car, cable, and general mounting 140
9.2   The technical procedure of logging 141
9.3   The logging truck 143
9.4   Drilling mud and logging 144
9.5   The logging systems 146
9.6   Electric logging systems 146
9.7   Nuclear logging systems 149
9.8   Acoustic logging systems 150
9.9   Some other types of logs 151
9.10   Stratigraphic correlation of bore hole logs 151
9.11   How to trace a fault or a disconformity by bore hole logs 152
9.12   How to evaluate a pay from bore hole logs 153

10.   *Open hole testing* 156

11.   *What a geologist should know about well completion* 162

Literature 166

Index 175

1 History of Petroleum and Petroleum Geology

Petroleum and natural gas belong to the minerals that have been used by humanity since the earliest ages, earlier than metals and coal, and for numerours different purposes. The peoples and tribes that found and used these useful but unusual materials gave them a lot of different names in their languages like: „Sweat of the Devil", „Oil from the Rocks", „Shining Water", and many others. Some of these names have survived thousands of years, e.g. „naphta", coming from the Babylonians and Assyrians, and „petroleum", derived from the Greek word „petros" for rock and the Roman word „oleum" for oil. The queerest name is that of „mumia", used by medieval physicians, because oil was used for many centuries to conserve mummies in Egyptian tombs.

Contacts between man and petroleum are so numerous and varied that several very interesting books have been written about these relations between humanity, history and petroleum in nearly every language and on every continent. Some of these books are listed at the end of this booklet.

1.1 The Use of Petroleum during the Ages

Reports on petroleum and especially on the use and production of petroleum have been found in Mesopotamian libraries, written in cuneiform characters about 4000 years B.C., in Egyptian pyramids written in hieroglyph characters, in Chinese books of the Ming Dynasty (1468-1644 A.D.), they are contained in Aztec books of Mexico before the Conquistadors, written in picture-writing on the bark of birches, and they are widespread in medieval European books, written by hand in characters called fractura. Since the invention of printing, booklets, pamphlets and books on petroleum have increased in number and contents in a manner that seems to be connected somehow with the production of crude. *R. J. Forbes* has spent a lot of time and work on collecting these ancient and historical documents, and getting them translated and assessed for their historical significance.

Other evidence on the use of natural hydrocarbons in bygone ages has been brought by archaelogists, digging out old towns, temples and tombs in the Near East. Again and again they have found things of the daily life of those ages made with the help of petroleum and asphalt, like, for example, the famous banner of Ur ($\approx$ 2600 B.C.), houses built from loam bricks using asphalt as mortar, and even the remains of the „Tower of Babel" made of such bricks and asphalt. Egyptian tombs have yielded mummies conserved by petroleum rich in polysulphides, and the Hieroglyphics along the walls of such a tomb told of an agreement with some

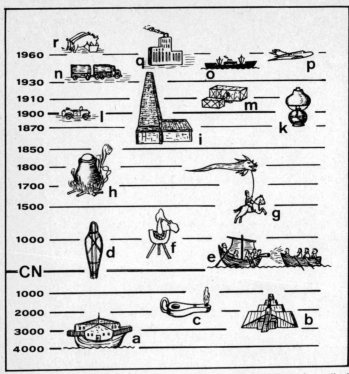

Fig. 1 The use of petroleum during the ages. a) The ark of Noah, caulked with tar. b) The temple of Ur-Nammu (Mesopotamia). The mortar between the bricks consists mainly of tar. c) An early oil lamp, from excavations in Mesopotamia. d) A mummy from Egypt, preserved with sulphuric tar. e) A warship, fighting with "greek fire". f) An early type of retort, used for the distillation of light fractions. g) A Mongol rider, using an oil lamp to light his dragon. h) Melting tar from bituminous rock (After Agricola). i) Drake's well, the begin of a new era. k) One of the first cheep kerosene lamps. l) An old-timer car. m) An aircraft from the 1st World War. n) An oldtimer lorry. o) Sea transport by motor ships. p) Air transport by modern planes. q) Production of electrical energy from oil and gas. r) Feeding pigs with yeast which has been produced from oil residues or from natural gas.

contractor on the shore of the Dead Sea in Palestine, who had to pay „royalties" for the production of such oil from that sea.

All these proofs, documents, old books and eshibits in museums are nothing compared with the enormous deluge of periodicals, literature and papers on petroleum and natural gas written every year since these natural hydrocarbons have become the most important source of energy all over the world.

1.1.1 Petroleum and Energy

Since the invention of machines people have been seeking energy to keep these machines running. Wind and water, horses, donkeys and even men had to move them until the steam engine was invented and coal was used to drive locomotives and machinery, especially in factories. The first big expansion of industry towards the end of the past century was the consequence of this development. It was followed by a second and still bigger expansion when electricity could be produced from coal and water in sufficient quantities to replace the big and noisy stationary steam engines by handy and nearly silent electric motors.

When gasoline and kerosene motors were invented in the first decade of the 20th century, this new kind of machinery enabled humanity to defeat space and time by cars, buses, planes and even missiles. Without any doubt petroleum has become the most important source of energy all over the world and is still increasing in importance relative to water and coal. Nuclear energy is coming into the energy market somewhat slowly in relation to what had been predicted, and up to now covers only a very small percentage of energy consumption. Even if enough electrical energy could be produced by nuclear processes, there will always remain a big market for petroleum, because it is the kind of energy that may be carried most easily even over long distances and stored wherever necessary at a minimum price. At any rate it should be mentioned that petroleum and its numerous products are really too valuable to be used mainly as a source of energy. Hydrocarbons are of nearly invaluable importance for chemical use. Solvents, plastics of every kind, resins and paints may be produced relatively easily using hydrocarbons as primary material, and the time seems to be near when even food may be produced in big quantities from methane or even residual oil.

1.1.2 Petroleum and Economy

Energy of every kind is the basis of modern industrial life and widely affects the economy. Petroleum plays an important role in the economy of producing and consuming countries since it has become the primary source of energy. Oil producing countries that were poor and nearly unknown such as Bahrain on the Persian Gulf sometimes became prosperous nearly overnight. In most countries the government keeps all existant and future hydrocarbon desposits under its control and so is in a position to use the profit coming from this source of welfare. In some of those nations the conditions of life have improved considerably. Hospitals, schools, water supplies and streets have been built, taxes have been reduced and credits have been given to build up new industries.

Other countries have enlarged and modernised their army, built new and tremendous palaces for their nobility, or transferred enormous sums of money to banks in Switzerland or in Beirut.

Most petroleum is not produced by the landowning countries but by international companies which have to pay „royalties" for their activities. These royalties recently have been increased by the countries owning the land and are a constant source of friction between these countries and the producing companies. Several nations have even nationalised their petroleum industry, expelling the foreign companies who found the oil fields. In most cases they had to call them back or to ask for help from other countries to keep the production going, because oil exploration and oil production need a complicated system of supply industries, geological, geophysical and technical specialists and experienced crews in order to succeed.

1.1.3 Petroleum and Illumination

For several thousand years medium and lighter hydrocarbons have been used for illumination. Small oil lamps made from earthenware are found in different forms in nearly all countries, and very often petroleum has been used for fuel. Especially in Mesopotamia and similar petroleum producing countries the medium fractions were

Fig. 2 Petroleum and illumination. An early oil lamp from Mesopotamia, a Roman chandelier, a kerosene lamp of our grandfather's time, and a security lamp for coal miners.

used for burning and the illumination of houses and streets, heavy fractions for torches and open fire pots. Egyptian alchemists invented the distillation of crude and even a kind of cracking process. The medium fractions like kerosene were mostly used for illumination, because the danger of explosions was smaller. Production in those times was too small, however, to be of real economic interest.

This situation changed suddenly when *Drake* drilled his first well producing several cubic metres a day, and soon the oil producers had to look for new markets for their products. A cheap but effective type of kerosene lamp was invented and sold below cost price only in order to make people buy kerosene in bigger quantities. Even nowadays similar kerosene lamps are used in isolated houses and in villages without electrical energy. More effective types have been invented where kerosene or paraffin oil is burned under pressure and which give better light by an incandescent mantle. Everybody knows such lamps, such as those used by fishermen to bait fish at night. Others use LPG (liquid petroleum gases) like propane and butane, sold in bottles of steel or aluminium for illumination and cooking, and especially for camping.

In general the importance of hydrocarbons for illumination has become insignificant since the invention of electric light. On the other hand a big part of that electricity is produced from petroleum, though strenuous efforts are being made to replace it by nuclear energy.

1.1.4 Petroleum and War

During the earliest times petroleum may have been used for fire-brands and fire-darts only. Sometime after the classical era Greek alchemists discovered a most striking military use for light oils, the so called „Greek fire". These first incendiary bombs of the world consisted of petroleum, sulphur and similar materials. The mixture, contained in pots of earthenware or glass were thrown by hand and burst into flames when hitting water or moisture. *Africanus,* a philosopher of that time, gave a recipe for making such a bomb out of sulphur, tar, light oil, salt and quicklime. The heat of the chemical reaction between lime and water was supposed to ignite the mixture, so its most important use was against ships and towns. As long as these bombs were thrown by hand, their efficiency remained somewhat small. Later on they were thrown by catapults and projectors, and this weapon became really disastrous when the Greek technician *Ktsebios* invented a double working pump, which enabled a ship to destroy an enemy ship by a jet of fire. In the beginning the Greeks used this weapon against Arabs and Russians, later on the Arabs learned to use the invention themselves, burning Greek ships in their turn. After serious losses on all sides the use of this disastrous weapon was forbidden by the Council of the Lateran in 1139 A.D.

Less than 800 years later soldiers in the first World War used a kind of weapon called a flame-thrower, consisting of one steel bottle containing petroleum and another bottle containing compressed nitrogen. With flames several metres long they burned their adversaries, and were burned themselves. This time no council banned the use of this terrible weapon.

Fig. 3 Petroleum and war. A greek ship throwing "greek" fire against an enemy ship. (After a contemporanian picture, 826 A. D.)

During the second World War thousands of civilians were killed by incendiary bombs containing benzene and phosphorus.

The latest „success" of the war industry in that respect is the Napalm-bomb, containing metal powders besides petroleum and phosphorus, a really disastrous weapon.

On the other hand no modern war can be made without petroleum for the cars, lorries, tanks and planes, and so petroleum became linked to politics. Wars have been carried on in order to get possession of oil-producing regions, and wars have been lost because of insufficient petroleum supply. Even in times of peace hard political fights between oil-producing and oil-consuming nations are a common thing.

1.1.5 Petroleum and Pollution

Since the industrial nations are becoming aware of the danger of killing themselves by extreme and uncontrolled pollution, a lot of papers have been written on that subject and especially on the pollution caused by natural hydrocarbons.

A lot of research has been done, mostly by the petroleum companies themselves. In general, specialists have come to the conclusion that pollution by crude oil and petroleum products is relatively small compared with that caused by coal or nuclear plants.

Most complaints about pollution by oil concern pollution of the air by sulphuric gases and gaseous oxidation products. These sulphuric gases, mostly sulphur dioxide, polysulphides, and sometimes hydrogen sulphide come from suphuric crudes or natural gas. Millions of tons of sulphur are extracted from hydrocarbons every year, but of course it is a question of the price of that purification process how far it may be pushed forward.

Crude itself is relatively harmless as to pollution, and most reports of enormous oil spills on the sea are reports of mere accidents. Much more dangerous to life in the sea is the constant oil spill by tankers washing their tanks in the open sea. Avoiding such oil spills and fighting them by harmless chemicals and methods are merely a question of international laws and of a corresponding control to keep the seas clean.

Even more serious than such oil spills may be the accidental pollution of water supply systems by medium and light petroleum products. A tank car with a cargo of gasolene really may cause a lot of damage when the fuel flows into a river or lake, but even such bad accidents are not worse than spills of other chemicals and may be fought by suitable methods.

Modern methods of underground storage diminish the contamination of ground water and even fight the „pollution" of the landscape by a series of enormous tanks which would be necessary to store the same quantity of oil or oil products.

1.1.6 Petroleum and Medicine

Some types of petroleum contain compounds of sulphur which are antiseptic, i.e. which kill bacteria. Even in modern pharmacies many remedies made of such polysulphides are sold for diseases of the skin, frostbite, inflammation, scabies and other kinds of illness. This antiseptic effect of some types of petroleum, known by their bad smell, was already known to priests in Egypt at the time of the Pharaohs. They used to embalm the corpses of their kings with petroleum coming from a seepage in the Dead Sea. In earlier times they took expensive natural essential oils for this purpose, but beginning with the 22nd dynasty (from about 945-745 B.C.) they learned to use cheaper oil from Palestine. Physicians from Egypt and Arabia, the best physicians of the Old World in the early Middle Ages, took the same kind of liquid for treating external wounds and diseases, using the name „mumia" and sometimes taking this medicine even from old mummies stolen from their tombs. During the crusades (11-13 century A.D.) the knowledge and wisdom of these physicians was brought to Italy, where it was taught in the universities of Venice, Parma, Verona and others in northern Italy and later on in France and Germany. Together with the knowledge of antisepsis the first remedies against bacteria came to Europe, and among others the black smear called mumia, which sold at amazing prices. Because of its price it was often adulterated by the addition of vegetable oil, tar and such substances, in later times even by petroleum from local seepages. Such local deposits often got famous for medical purposes, and sovereigns and landowners carried on a flourishing trade with their petroleum production. The most famous and most valuable oils of that time were the so called oil of St. Catherine from Modena in Northern Italy, the St. Quirinus oil from a monastery in Bavaria, and the Thyrsus oil from Seefeld near Innsbruck in Austria.

Fig. 4 A medieval surgeon, treating the head wound of his patient with oil. (The Geneva pamphlet, 1490 A. D.)

The last mentioned oil is still produced today from bituminous rock by a small plant and sold all over the world as an external antiseptic medicine. All these oils for medical purposes were boosted in a manner quite alien to our pharmaceutical industry, and sold at exorbitant prices. Thus the so called Thyrsus oil was said to be the blood of a giant, who lost his arm in a fight for a good purpose, and was sold for one gold piece (about 10 dollars) a pint. On the other hand production at that time was very poor, even in relation to a bad stripper well of our times. From oil seeps near Hanover (Germany) we know that a skilful oilworker could produce about 1-2 litres of petroleum a day, at a price of about one mark a litre by scooping several cubic metres of water out of a deep tar pit.

Nowadays the pharmaceutical industry could live very well without petroleum and petroleum products, though a big number of medicines and pharmaceutical liquids are still produced from petroleum.

In general, and first and foremost, the packaging of pharmaceutical products now comes from petroleum rather than the contents, because most of the tubes and bottles consist of plastic material made from petroleum or natural gas.

1.1.7 Petroleum and Religion

Readers in general might think it difficult to find a real connection between such a bad smelling, dirty stuff like crude oil and a religion of any kind, but *Forbes* showed among other things that petroleum was in many cases linked with or at least mentioned by religions in countries where it occurred.

The oldest known building made with the help of asphalt is a bath in the valley of the Indus river, built about 4000 years B.C., and without doubt devoted to ritual bathing.

The oldest document mentioning petroleum deals with the first man, called *Ut-Napischti* and was found in Mesopotamia carved in cuneiform letters on small earthen plates. One part of this tale is the equivalent of the story of Noah in our Bible and seems to be a primary source of the tale of the great flood. In Mesopotamia, such deluges were and are common events, so the description how *Ut-Napischti* built his ark sounds like a practical instruction for the shipbuilding of that time. Like all shipbuilders that make their ships from wood, *Ut-Napischti* took asphalt to caulk his ark, and probably asphalt from tar pits on the Euphrates river, not too far from the modern oilfields.

A very close connection has always existed between petroleum and the religion of the fireworshippers (Parsees) in Persia and other countries of the Near and Middle East. Their name comes from the eternal fire they have in their temples. The priests of this religion have to keep the fire burning, and if possible they use petroleum or natural gas for this purpose. Travellers of older times reported of such temples with fires fed by natural gas coming from seepages several hundred metres away through leaden pipes. So the priests of this religion seem to be the first men able to complete a gas well by casing and wellhead and to build pipelines for gas, several thousand years before our time.

Even the Bible in its older parts is full of tales about petroleum. The Jews of that time lived in a country famous for its petroleum deposits, and in nearly every chapter there are tales of events that could be either of volcanic origin or caused by wild eruptions of

Fig. 5 A temple of the Fire-Worshippers. (After a paper of Thomas Hyde, 1760 A. D.)

Fig. 6 An eternal lamp fed by a leaden pipeline from a subterraneous reservoir. (After Athanasius Kirchner, 1682 A. D.)

petroleum or natural gases. The most striking event is that of Sodom and Gomorrha, two towns on the shore of the Dead Sea which were destroyed by a catastrophe accompanied by an eruption of sulphur hydrocarbon gases. Greek philosophers later on, e.g. *Strabo,* described similar and smaller eruptions of a heavy crude in the same region, some centuries after that report in the first book of the Genesis.

Other tales even mention incidents with burning petroleum, killing some Chaldean men, and of a burning rock which burned some gifts laid on it. It would be of maximum interest to know more about these relations between petroleum and religion. This would be very intersting for sociologists and theologists, but is a bit out of place in a geologic text book.

1.1.8 Petroleum and Food

One of the most modern branches of petroleum industry is the production of food from oil or natural gas. Some races of yeast live on oil residues, others are able to decompose methane. Yeast on the other hand is a valuable protein food which may be fed to chickens and pigs. Production of yeast protein from natural gas is quite simple in principle. The yeast cells are kempt in a solution of some potash, nitric and phosphoric salts in water and the gas will stream through the water. Surplus yeast cells may be filtrated off, dried and sold to chicken or pig breeders. Pilot plants using this principle are said to work well at a reasonable price for the product. Countries with big gas fields and a constant lack of protein, such as Algeria, could easily fight this shortage of high quality food.

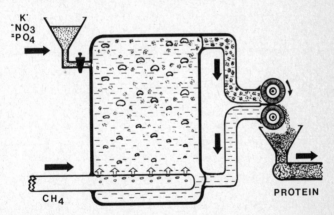

Fig. 7 Principle of proteine production from natural gas. The natural gas and some nutrition salts are fed into a water tank, where the yeast cells grow. The suspension of yeast is filtered off and may be fed to pigs or chicken.

Some difficulties still seem to exist with regard to the other principle, using heavy residues of oil rich in paraffin for basic material. Most of these substances have a very bad smell, coming from sulphuric compounds, and this bad smell even clings to the pig meat produced from such oil. Anyhow this divergent use of petroleum hydrocarbons will certainly lead to a considerable production of food, and especially of proteins.

1.2   Petroleum and the Future of this Kind of Energy

Nearly all authors writing about petroleum and its future used to speak of reserves that will not exceed 30 years or so. Quite often politicians and people interested in the future of energy supply wonder what will happen at the end of this period. This kind of predicting reserves is nearly as old as oil industry, and even at the end of the last century reports on oil reserves show the same prediction of about 30 years. In reality there will be oil and gas for much a longer time, and every oilman will only smile when he reads comments on such misunderstood predictions.

The reason for this apparent contradiction is based on the method of calculation:

Every oilwell, after drilling and completion, has a restricted time of production, in average about 30 years. That is the reason why even oilfields·have a restricted time of life, i.e. in principle not longer than the last well that has been drilled and exhausted, and so in general 30 years. At any moment when such a prediction is made, it can only be made comprising all oilfields and oilwells in being at that time, with their general maximum lifetime of about 30 years. Of course everybody knows and hopes that new structures and oilfields will be found as soon as possible, but the crust of our earth is not transparent enough to allow any prediction on quantities which we cannot see or estimate or calculate but only hope to be existent.

The history of oil exploration has shown that there will always be enough possibilities of finding oil and gas in sufficient quantities if enough exploration is done in the right places, but if you ask a specialist, he will always say that our reserves on crude and gas will only be sufficient for another 30 years.

1.3   History of Petroleum Exploration and of Petroleum Geology

So long as oilmen knew nothing about the origin, the nature and the accumulation of oil and natural gas, every one of their oil finds was just a matter of chance and accident. Natural seepages, oil seeps and gas volcanoes provided the first contact between petroleum and humanity and remained the only connection for several thousand years. Perhaps the priests of the Parsees knew some principles of exploration and exploitation, but they kept their knowledge secret.

When petroleum became an industrial factor during the end of the last century, oil exploration began from these old and wellknown natural occurrences. In many countries shafts were sunk and holes were drilled near old seepages. Many lucky finders got rich overnight when they struck a good pay, or ran into financial ruin when an expensive borehole turned out to be dry, or if some accident of drilling made it impossible to finish the well.

Theories about the laws of oil accumulation for a long time remained on the whole mere fantasy and misunderstood experience.

Thus the so called line theory governed the thoughts of German petroleum geologists for a long time. At that time people used to connect all known occurrences of oil by lines on a map, and other occurrences were expected to be situated somewhere on these lines and in some more or less orderly arrangement between the known fields. At that time nearly all oilfields were on the flanks of or near salt domes, and these salt domes are grouped along the directions of two tectonic systems crossing one another. So the experience about these lines was right in principle, but led to negative results when applied to other regions. Only later geophysicists learned to find salt domes in the underground by means of gravimetric and seismic surveys.

On the other hand, one of these prime theories proved to be correct in principle, the so called anticlinal theory.

This theory came from Pennsylvania where geologists discovered that oil occurred mostly in the highest parts of sandstones. Such higher parts of a fold are called anticline in geology, and this principle is still valid and has remained the basis of all geological and geophysical methods when searching for oil and gas.

Such anticlines are rarely visible from the surface, and geophysics began to play their part in the growing petroleum industry. Gravimetric, magnetic and geoelectric methods were applied, improved and developed to fit the needs of oil exploration. The most striking progress was the invention of refraction seismics and later on of reflection seismics in the early twenties of this century. Every real improvement of these methods resulted in the detection of series of new finds. Even during the last few years new systems have been invented, comprising exploration by so called remote sensing from planes, helicopters and from satellites.

Not only geophysical but also geological methods had to be improved to fit the requirements of drilling site and oil field geology. The necessity to date minute brickles of sediments created a completely new technique called micropaleontology, which multiplied the number of known fossils by four or six. Analysis of sediments, of sandstones and of porous limestones were carried out in order to know more about the structure of the pay and the movement of fluids in its pore space. Theoretical results from laboratories had to be applied to the field-work and could be controlled by the results of production. One of the most important methods invented in the geological sector proved to be the so called borehole logging, in principle the inversion of surface geophysical methods into borehole analysis. An entire industry was created from minute beginnings which soon brought in several billion dollars a year. These methods, based on electrical, nucleonic and acoustic principles enable geologists to describe entire stratigraphic series exactly to the inch. They make it possible to correlate one borehole with others in the vicinity or even hundreds of miles away. The exact and complex data logged by these methods in a pay zone initiated the development of a technique called production engineering, comprising a combination of petrophysics, geology and production technique. All that, or nearly all that, grew over a period of less than fifty years, an amazing combination of methods and theories taken from nearly all parts of natural sciences, and even from economy, sociology and juristic sciences.

On the other hand these complex methods of analysis produced enormous quantities of data from every well, and there are single oilfields containing more than a thousand wells. Storing, assorting, controlling, reworking and retrieval of so many data in a reasonable way cannot be done laboriously by hand. So even small and middle sized oil companies own or share a computer with a special library of software to do such work in a reasonable time and without error.

Constant physical controls of producing wells, fed by radio or telephone lines into such computers and reworked by process control programs enable companies to run fields – with all installations for treating and separation without any human involvement in the process.

2 Origin and Accumulation of Petroleum and Natural Gases

Various theories and suppositions have been published in the course of time about the origin of these hydrocarbons and their accumulation. It was a big step forward when it became a general opinion that origin and accumulation are two completely different things that have nothing to do with one another, neither in time nor in location. Petroleum and natural gases in smallest droplets or bubbles may be carried by flowing ground water over enormous distances within the pore space of sandstones or porous limestones. Nobody ever found a dependable way to determine the location or even the formation where this or that type of oil or gas comes from. Modern techniques of isotope analysis only enable specialists to date whether a particular gas originates from petroleum or from coal by their different $C12/C13$ quotient.

2.1 Theories about the Origin of Petroleum Hydrocarbons

There are two main groups of theories about the origin of these substances, the inorganic and the organic one.

The inorganic in general is the older one and simply says that petrol hydrocarbons originate from inorganic sources, such as volcanoes, carbides of metals, or simply by the reaction of carbon dioxide, water and catalysts with sunlight.

In most of these theories the evidence is provided by methane, which is found in volcanic gases, within the atmosphere of stars, in metamorphic rock and in other places where nobody would expect it to occur. Of course granites or other crystalline rocks in deeper layers of the earth may and will contain carbides which can create methane or similar gases by their reaction with water, but in general these theories have nothing to do with real petrol hydrocarbons. Methane is a species of gas that may originate from the most various organic and inorganic processes, and so is not good evidence for theories of inorganic origin. Other arguments came from countries where oil, asphalt and petrol tars occur in regions where no sediments but only metamorphic rock occur such as in central Sweden. Isotopic analysis of these oil tars however showed that they are of mesozoic age and must come from mesozoic sediments that were lying above the metamorphic rocks before. Meantime they have been eroded, leaving behind them only these rare drops and patches of bitumen. In general all these theories about the inorganic origin are now merely of historical interest only, but again and again somewhere in the world – usually a chemist and only rarely a geologist – finds new evidence or new aspects of old evidence for this principle.

The organic theories try to prove that the so-called petrobitumina originate from organic materials, and especially and mostly from planctonic organisms of the sea, of brackish waters and even from lakes. These organisms, living in such waters in enormous quantities,

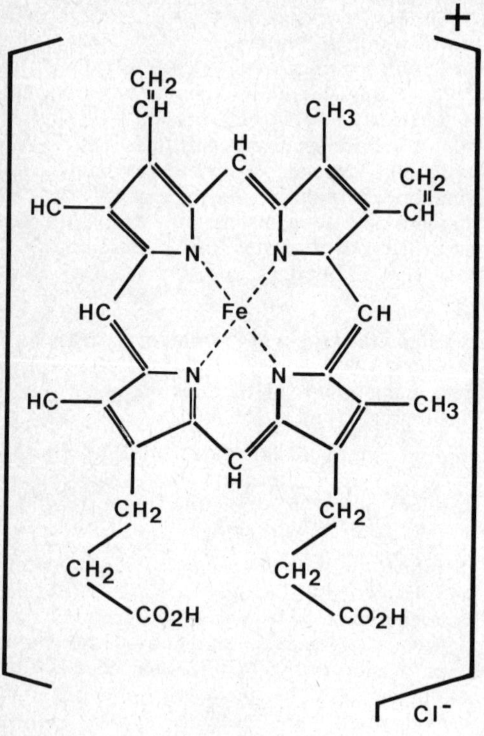

HEMIN

Fig. 8 The chemical structure of hemin, the active part of hemoglobine contained in the blood of most animals. Porphyrine rings like those surrounding the iron atom in the middle are found in crude oils. Such porphyrine rings would be destroyed at temperatures higher than 300 centigrades.

contain fatty acids and amino acids which are said to be the basic material of oil formation. Most of this plancton production occurs on the shelf around the continents and in waters not deeper than one thousand metres. Such organisms need phosphoric and nitric salts for their life. These salts come from onshore and nearshore

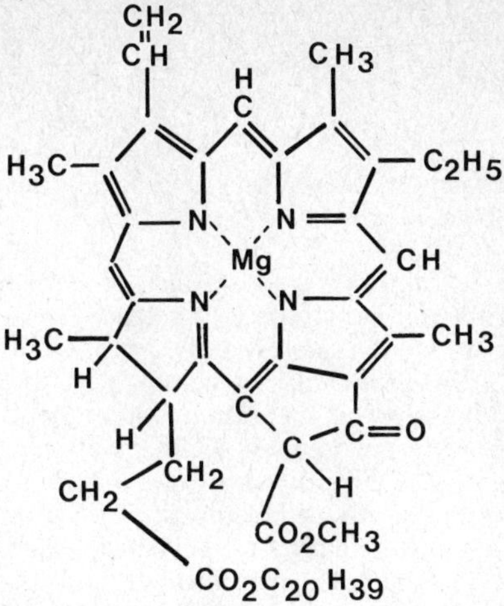

CHLOROPHYLL

Fig. 9 The chemical structure of chlorophyll, contained in most plants. Porphyrine rings like those surrounding the magnesium atom are found in crude oils.

sediments and are swept into the sea by rivers and estuaries. Part of these organisms are plants, mostly algae, and contain chlorophyll. This chemical compound belongs to the group of porphyrines and enables plants to build up complicated hydrocarbons from carbon dioxide and the energy of sunlight. Fragments of such porphyrine rings, deriving from chlorophyll, are found in all types of crude oil besides other fragments of haemoglobin and other similar compounds coming from the blood of animals, such as foraminifera, minute crustaceans and small worms. Iron oxides, copper and vanadium are found regularly in crudes and are said to come from the blood of animals where they form an important part within the porphyrine rings for the exchange of oxygen.

The production of plancton in the world and in some parts of the sea is really enormous. Within the Black Sea every year 2.7 billion tons of plancton are produced, containing on average 1-3 % of fatty acids and 4-16 % of amino acids.

These plancton organisms mostly live only a very short time. When they die, they are eaten by other organisms, or they sink to the bottom, where they become the prey of microbes which destroy their bodies and modify the chemical compounds contained in their tissues. In waters rich in oxygen all material which could be important for the genesis of oil can be oxidized and get lost, so that the preservation of oil and high content of oxygen are somehow incompatible.

Even in well-aerated waters, however, dead plancton on the bottom may be preserved from oxidization by sediment falling on it and blocking it off from oxygen. Generally in all sediments deposited in water there is a constant lack of oxygen. Even coarse sandstones in fresh waters only have a relately thin oxidation zone and show dark reduction zones only a few centimetres below their surface. So the probability of conservation is somehow high in all regions with active sedimentation, where organic material on the bottom is quickly covered by sediment and thus saved from oxidation. Within the sediment the organic material is overcome by a process that is commonly called the ripening process and which is not known in detail. Chemists usually speak of merely chemical processes, some sedimentologists speak of absorption and alternation by surface activity of clay particles, and some bacteriologists try to emphasise the role of microbes in this transformation.

Probably all these processes play a certain part in the transformation of dead organic material into petroleum hydrocarbons. Of course the resulting minute droplets of oil are only one product of this process. A certain part of the material will remain within the source rock for ever, and other parts will get into solution and be swept away with the interstitial water. Long series of chemical water analysis in various regions of the world have shown that underground waters bordering oilfields or near oilfields often contain typical chemicals, especially bromine and iodine, which probably may be derived from such processes of decomposition. The minute droplets of petroleum trapped by the sediment together with genuine sea water remain only for some time within their parent sediment. Such young sediments, especially young types of clay, contain many times more water than clays and claystones which we know from claypits or brickyards. They still have to be compacted and dehydrated to become a solid sediment. With this process they shrink to 1/20th or even to 1/100th of their original volume, giving off most of their original water to overlying sediments.

2.2 The Migration of Petroleum Hydrocarbons

All pores, caverns, and cavities in rock or sediments below the water table are filled with some liquid or gas, more than 99 % with water. The rest, restricted to very rare cases, may contain oil or gas. All these pore fillings are affected by a system of static and dynamic pressures which are the main reason for the migration and the trapping of petroleum hydrocarbons in reservoirs.

Most obvious is the pressure caused by the weight of overlying sediments, called overburden pressure. Sediments on an average have a specific weight of 2.5 to 3.0. At a depth of e.g. 1000 metres the sandstone forming them is compressed by a weight of 2.5 to 3.0 tons per square metre. All sediments have to give way to that overburden pressure as much as they can. Mudstones become compacted, losing a big part of their water content, and become solid claystones. Sandstones with a primer pore space of more than 40 percent lose part of their porosity down to 20 percent and less. The water squeezed out of the pore space moves towards higher zones with lower pressure. This compaction will cease only when the sandstone or claystone is able to carry the weight of the overlying sediment by its internal stability. Then the internal tension or pressure of the solid framwork equals the overburden pressure. If the pore space were empty, it would be under atmospheric pressure only. In reality the weight of the pore filling – mostly salt water – gives rise to a so-called formation pressure. Usually this formation pressure may be calculated by multiplying the depth in metres by a factor of 0.11. Regions with very young and unconsolidated sediments mostly show higher pressure gradients.

Other factors affecting the contents of the pore space are surface tensions between the surfaces of the sand grains, the water content, oil droplets and gas bubbles. During compaction the water surplus of the sediments will be expelled and move to zones of lower pressure, carrying with it droplets of oil and bubbles of gas. Under normal conditions water shows higher affinity to the rock surface; the pore space then is water wet and will always retain a fine film of water covering the surface of the rock, the so-called connate water.

Compaction of sediments in a basin which is sinking down will remain a permanent process over geologic times. The squeezed out water is kept moving for the same time and even longer, and the path of its migration may cover hundreds of kilometres, always within the narrow pore space and at a very low, hardly measurable speed. Many of these migration paths may lead into the open, and the oil droplets will be lost by oxidation. Others will be ended by so called traps or reservoirs, forming oil or gas fields and providing a foundation for the welfare of entire peoples and regions.

Before such an accumulation of petroleum hydrocarbons may be called a reservoir, oil droplets have to join to form bigger drops, and gas bubbles have to fill the gas cap of the field. It my take millions of years to fill such a reservoir with a considerable amount of hydrocarbons, but it takes only thirty years or little more to pump the oil out of a newly-found field.

2.3   Some Words about the Chemistry of Petroleum Hydrocarbons

Petroleum hydrocarbons or petrobitumina include an amazing variety of gases, liquids and even solid substances. Some of them are colourless and tasteless, others are brown or even black and really bad smelling. Only very rarely fields or pays contain one pure compound only, like some gas fields contain only methane. Most pays contain a complicated mixture of dozens or even hundreds of different compounds. A crude oil from the Ponca City field in Oklahoma has been analysed thoroughly. After more than 40 years of work, and after spending more than one million dollars, more than 230 different substances could be identified. These substances contain not only hydrogen and carbon, but also oxygen, nitrogen, sulphur, and even metals. Part of these elements come from the real source material of oil and gas, others have been absorbed from pore fluids or from surrounding rock. Systematically the petroleum hydrocarbons or petrobitumina belong to the group of natural hydrocarbons, which in total include the petrobitumina. The petrobitumina comprise natural gases, condensates, crude oils, ozokerite and asphaltites. They are able to migrate and are soluble in carbon disulphide.

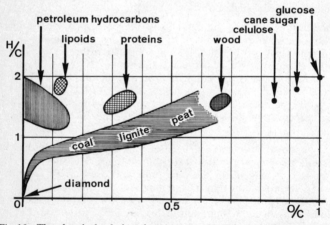

Fig. 10 The chemical relations between petroleum hydrocarbons and other natural hydrocarbons, shown by their ratio of oxygen versus carbon, plotted against the ratio of hydrogen versus carbon.

The kerobitumina comprise oillike substances like those oils which occur in bituminous slates and bituminous limestones. They are not able to migrate, and are not soluble in carbon disulphide.

A really big and important group is that comprising the carbo-bitumina, peat, lignite, coal, and anthracite.

Boghead coal, fossil waxes and fossil resins are of uncertain position.

Methane, for instance, may originate either from the decomposition of plants during the diagenesis of coal or from processes accompanying the genesis or the decomposition of petroleum hydrocarbons.

Only since a few years it is possible to state the origin of such gases with some certainty by their $^{12}C/\ ^{13}C$ ratio. These investigations show that the enormous gas fields in Holland and northwestern Germany originate from the coal bearing Upper Carboniferous in their underground.

Graphite and diamonds are other extreme ends of these diagenetic series. Their exact origin is unknown, and it is really marvellous that a smeary and stinking stuff like crude oil may become a clear and shining diamond not only as a result of financial transactions but even by geologic metamorphosis.

All those natural petroleum gases, condensates, crude oils and asphaltites may contain only four different groups of hydrocarbons:
The alkanes or paraffins,
The cycloalkanes or naphthenes,
the aromates, and complex hydrocarbons.

The structure of alkane compounds show a linear arrangement of carbon atoms with or without lateral branches, according to the general formula C_nH_{2n+2}, beginning with methane and ending with solid paraffin. With normal temperature and pressure, the compounds from 1 to 5 C – atoms are gaseous, those up to 16 C – atoms are liquid, and the others solid. Under reservoir conditions this may be different, especially with the so-called condensates, from C_3 to C_6. Their state of aggregation varies – according to the counteraction of temperature and pressure – within the pay, the well and within pipelines or tanks. All alkanes are colourless, tasteless and belong to the agreeable substances in crudes, as long as the chains are not too long.

Cycloalkanes or naphtenic compounds show a cyclic arrangement of the carbon atoms with only single valences connecting the carbon atoms, so-called saturated compounds. Most important are the liquid compounds Cyclopentan and Cyclohexan. As long as these substances are monocyclic and without lateral branches, they may be represented by the general formula (C_nH_{2n}). They may have sidebranches, and even become bicyclic like decaline.

Aromates are unsaturated cyclic compounds with partly double connections of the carbon atoms. They are mostly liquids and have very characteristics odours, like benzole, toluol or xylol.

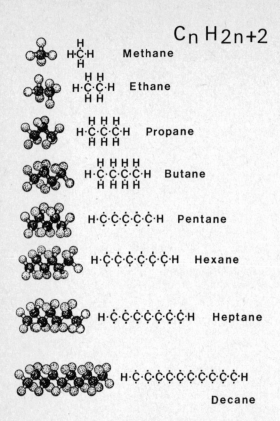

$$C_n H_{2n+2}$$

Methane

Ethane

Propane

Butane

Pentane

Hexane

Heptane

Decane

Fig. 11 The chemical structure and composition of some alcanes or paraffins. The smaller chains from methane till to pentane are gases (at normal temperature and pressure). The longer chains till to 16 atoms of carbons are liquids. Alcanes with more than 16 atoms of carbon are more or less solid.

Other compounds of this group have lateral branches or are polycyclic. Polycyclic compounds are relatively rich in carbon and poor in hydrogen. Complex hydrocarbons contain aromatic, unsaturated rings as naphthenic, as well as saturated ones and straight or branched alcanic chains. Other authors call them naphtenic aromates or hybrid hydrocarbons. One of such compounds is tetralin ($C_{10} H_{12}$) with one naphthenic and one aromatic ring.

Sulphuric compounds are somehow common in crudes and natural gases. They may originate from the protein of plants or animals or come from the reduction of gypsum dissolved in the formation water. This reduction is effected by bacteria of the genus

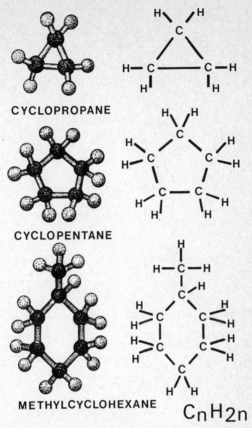

CYCLOPROPANE

CYCLOPENTANE

METHYLCYCLOHEXANE

C_nH_{2n}

Fig. 12 The chemical structure and composition of some naphthenes.

Desulfovibrio, which decompose sulphates and hydrocarbons and transform them into a porous kind of limestone and hydrogen sulphide or solid sulphur.

$$C_xH_x + CaSO_4 + xH_2O \quad CaCO_3 + H_2S \text{ (or S)} + xH_2O$$

These bacteria occur in nearly all sediments and seem to have their function best in the caprock formation of salt domes which are not too deep below the surface, because they remain active only at temperatures below $60°C$.[Caprock and other sediments containing sulphur in layers deeper than that temperature must have been in higher position during the genesis of that sulphur.]

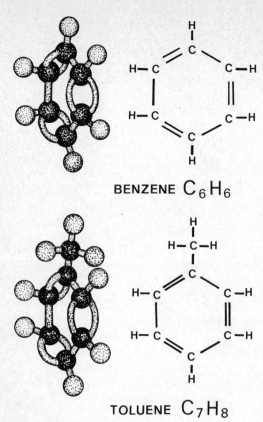

BENZENE C_6H_6

TOLUENE C_7H_8

Fig. 13 The chemical structure and composition of two common aromates, benzene and toluene.

Sour gases and crudes rich in sulphur compounds, therefore, are abundant near salt domes or formations rich in gypsum or anhydrite. They may contain solid sulphur, hydrogen sulphide, or sulphuric compounds like mercaptane, disulphide, sulphides, or thiophene. All these compounds have strong and bad odours, are mostly poisonous and tend to cause technical and chemical difficulties. Therefore they have to be extracted from gas or crude by special stripping and washing processes before the hydrocarbons may be treated in refining plants by catalytic processes or sold for consumption. Some very sour gases contain more than 10 percent of hydrogen sulphide which must be extracted before the gas can be fed into a pipeline or burnt. Gas- and oil-fields therefore have become the most important sources of sulphur supply. The gas-field of Lacq in southern France, for example, produces more than 600,000 t/year of sulphur, more than all the sulphur mines of Sicily during their best times of production.

Oxygen compounds may occur in naphthenic acids, fatty acids and compounds derived from phenol. They form by oxidization of these substances by atmospheric oxygen and form mostly resins which alternate the fluid properties of crude by an increase of viscosity and by their surface activity. They may change carbonaceous pays from water wet to oil wet property.

Nitrogen compounds are not sour like the above-mentioned resins but are mostly basic. They are the reason for the black colour of asphalt and asphaltic crudes and are for the most part chemically unstable.

2.4   The Pay and its Properties

A pay is a porous and permeable formation which is able to gather and to produce petroleum hydrocarbons. It may, for example, be a

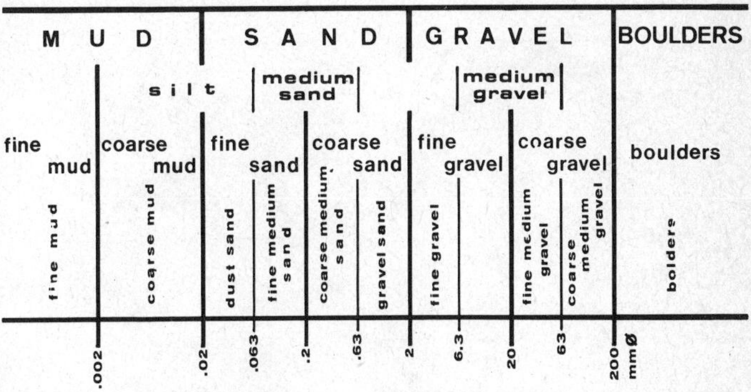

Fig. 14   One of the various systems for granulometrical analysis of clastic sediments.

sand or sandstone, an oolith, a vugular or cavernous limestone or even fissured crystalline or volcanic rock. Porosity and permeability may be original, dating from their sedimentation, or of secondary origin, caused by diagenesis or tectonics. Some pays are completely tight with respect to their type of rock, and their oil content is accumulated in fractures and fissures only. Others have a certain porosity, but their permeability is too low for a good 'production. They produce their oil or gas by means or such fractures or fissures. On the other hand it is a common practice of production engineers to stimulate wells with low permeability pays by hydraulic or chemical fracturing.

A good pay for producing oil should have a porosity higher than 20 percent and a permeability of more than 300 millidarcy. For producing gas there are lower requirements. There are gas fields with a relatively good production which only show a porosity of 10 to 11 percent and a permeability of only 25 to 40 millidarcy. For comparison, a beach sand may have a maximum prosity of about 45 percent and a permeability of more than 6 or 8 Darcy.

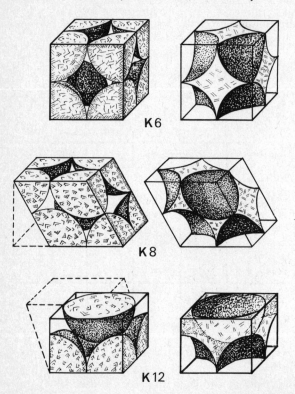

Fig. 15 Round grains of equal size, packed together according to different systems. The uppermost couple of models shows the smallest number of contacts between the grains (K 6) and the biggest pore space. The lowermost couple shows the highest number of contacts (K 12) and the smallest pore space. (After v. Engelhardt)

Fig. 16 A model of round grains of equal size, packed together at a rate of 12 contacts per grain.

Fig. 17 The pore space of the same model. The grain size itself will not affect the amount of porosity.

The easiest way to understand the properties of a pay, the accumulation and production of oil and gas, and the possibilities of investigating such properties in bore holes and on cores, is to inspect granular pays like sands, sandstones and ooliths, or even models made of balls. If a number of balls of the same size is shaken in a container, they will – after some time – reach a state of maximum density, where all balls touch one another in more than 10 single points, with a maximum of 12 points. In this state of maximum density they show a porosity, i.e. the part of the container filled with air, of little more than 40 percent. The size of the balls does not matter, little balls of equal size show the same porosity as big ones.

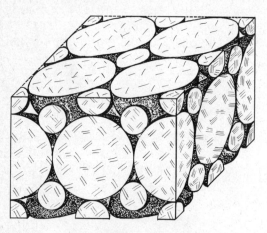

Fig. 18   Model of a packing of grains of two different sizes. The smaller grains fill part of the pore space between the bigger grains.

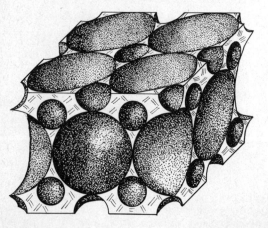

Fig. 19   Model of the pore space in fig. 18. This pore space is much smaller than that of fig. 17.

If a mixture of big and small balls is fed into the container, however, the small balls may fill part of the former empty pore space and reduce porosity. The crucial thing for the porosity of granular pays therefore is their combination of grain sizes, investigated by a grain size analysis. That may be done in the field by aid of a measuring microscope provided with a scale, or in the laboratory by crushing a sample and washing it through a set of sieves with screens of different size. The percentage of the sample remaining on each sieve may be represented by a frequency curve, or better by a summation frequency curve. Such a curve, represented on a probability grid, will show a straight vertical line for a sand of equal grain size, and straight but oblique lines for usual mixtures of different grain sizes. From the obliqueness of such curves the porosity may be calculated with some probability. Even the interior surface may be calculated from such a grain size analysis, if it is not measured directly by a special apparatus. Some sandstones show remarkable differences between porosity dates calculated from a grain size analysis and dates derived from bore hole logging or laboratory porosity tests by weighting the dry sample in a fluid of known density.

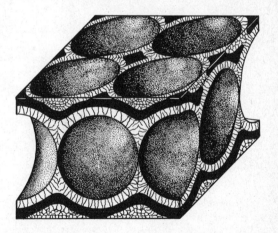

Fig. 20 The influence of cementation. The grain model resembles that of fig. 16 und 17, but part of the pore space is filled with carbonaceous or siliceous material. The remaining pore space is much smaller than the original space.

Such sandstones usually show under the microscope a more or less conspicuous cementation with carbonaceous, siliceous or clayey material. This cementation of the pore space is a diagenetic process which usually reduces the rate of porosity by filling part of the pore space or even dosing off part of it. So we speak of dead pore space, which cannot be used for production, of effective pore space, which is open for production, and of bulk porosity, comprising both types of pore space.

The permeability of a pay is represented by the total of all finer and bigger pores, which take part in flowing processes within the pay. It may be calculated from a grain size analysis with some probability, measured on cores in a laboratory, or calculated from the production rate of the well. Coarse sands and sandstones always

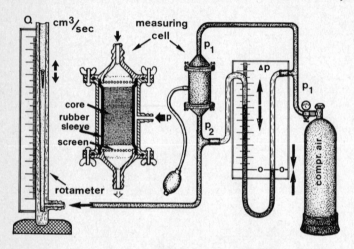

Fig. 21 One of the methods to measure the permeability of a core. The small core (plug) is fed into a special cell. A rubber sleeve keeps the sides of the plug sealed. Air is pumped through the plug, and the amount of air versus time is measured with a rotameter or a bubble counter. The differential pressure before and behind the cell may be read from a manometer.

have a higher permeability than fine grained ones, of course. Most pays show striking differences in permeability from one layer to another. If such fields are produced too rapidly, they will produce only from the highly permeable layers and oil within the finer grained streaks will be cut off by the edge water and be lost for production. Care should be taken in such cases to open the worst parts of the pay first by perforation, or to improve their flowing properties by suitable hydraulic or chemical fracturing.

The most important thing for production, however, is the thickness of the pay. There are sandstones and sands with a thickness of more than a hundred metres, and other of only a few centimetres. Pays with impermeable intercalations of shale, lime or similar rock are called multiple pays, and if these intercalations are thin and

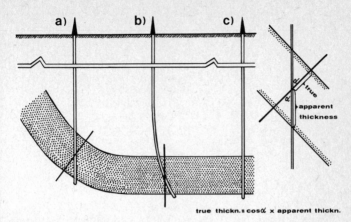

true thickn.≈cosα x apparent thickn.

Fig. 22 Three wells penetrating the same pay, showing different (apparent) thickness. a) The well is perpendicular, the pay is not horizontal. The true thickness must be calculated. b) The pay is horizontal, but the well shows a certain dip. The procedure of calculation is similar to the preceding one. c) The well is perpendicular, and the pay is horizontal. Apparent thickness and true thickness are identical.

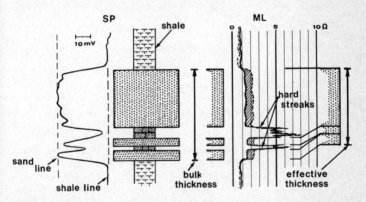

Fig. 23 A pay with some hard, impervious streaks between porous rock. The effective thickness of the pay may be calculated from the logs, or may be derived from the cores.

numerous, geologists speak of a sandwich formation. In such cases a difference must be made between the total or bulk thickness of the pay and its effective thickness, which means the total thickness of all permeable streaks and layers. If a pay is hit by a well, and neither the pay is horizontal nor the bore hole really vertical, the thickness open to production is called apparent thickness and may be reduced to true thickness by a graphic plot or trigonometric calculation.

When a field is developed, the pay is investigated in detail and in total. The most important results of such investigations are represented in structural or contour line plans presenting the top and the base of the pay, its cumulative, effective and total thickness, and other properties like average permeability, average porosity and the quantity of „oil in place" per square metre. Such plans provide as the main basis for the further development of the field, for controlling production, and for initiating secondary and tertiary methods of production.

2.5   The Structures and Contents of Oil- and Gas-Bearing Reservoirs

Gas- or oil-fields represent the temporary end of the way of migration, temporary in the geologic meaning of that term. The development of such a reservoir may take several million years, and it may remain intact over long geologic periods. Tectonic movements can destroy it, opening new ways of migration, and heat and pressure in deeper strata of the earth's crust may transform gaseous or liquid hydrocarbons into solid asphalt or even into graphite.

Quite a number of conditions must be met before such a reservoir may develop. In the first place there must be a sedimentary basin with enough mother rock to supply sufficient quantities of petroleum hydrocarbons. There must be permeable layers in the right position to facilitate migration, there must be a suitable porous and permeable pay formations, and last but not least there must be a so called trap, where the way of migration is interrupted and the hydrocarbons are accumulated. Such traps are more or less complicated tectonic, sedimentologic or stratigraphic structures with impervious boundaries in the right places. Materials which are practically impervious to oil and gas are, for example, clay stones, marls, tight carbonates, rock salt, anhydrite and previously porous sediments with impermeable pore fillings. These impermeable boundaries have to interrupt the way of migration, i.e. they have to shut off the permeable pay formation against higher zones of lower pressure. Within the trap the droplets of oil and the bubbles of gas accumulate in the pore space of the pay. The petroleum hydrocarbons segregate according to their different gravity and form normally a gas cap on top, an oilbearing formation below over a basis of formation water. The boundaries are horizontal in general, but not a sharp as one would think, and this entire scheme in reality is much more complicated than is seems to be.

2.5.1 What Does the Pore Space of the Pay Contain in Reality?

The primer pore filling of all marine, brackish or limnic sediments is water with a varying amount of dissolved salts. Even within the gas cap and the oil-bearing formation a certain amount of water remains within the pore space: the so called connate water. It covers the walls of the pore space like a very fine film and is thicker where ever two grains touch one another. The percentage of connate water in a pay depends on the dimension of the inner surface, and this depends on the average grain size. Fine-grained pays may have a percentage of connate water up to 70%, but may produce oil or gas. Normal sandstones with medium grain size will have about 15-20 percent of connate water.

In gas fields the gas-bearing formation containing gas and connate water in its pore space lies above edge or bottom water with only water in the pore space. Under reservoir conditions, i.e. under the pressure and the temperature of the reservoir some gas is dissolved within the edge water: at a depth of 1000 metres about 6 cubic metres of gas in one cubic metre of water. The transition zone from the gas-bearing formation to the zone of water with dissolved gas is relatively small, one or two metres thick. Its thickness depends mostly on the type of pay and its grain size. If a well hits the transition zone it will produce mostly gas, because gas has a lower viscosity and a higher mobility in the pore space than water. If a well hits the pay just outside the transition zone, it will produce only water with bubbles of gas, which may go out of solution so abruptly that they cause minor blow outs.

Oil fields show a very complicated vertical distribution of various pore fillings when they are studied in detail. Within the so called gas cap the pore space contains gas and connate water like a gas field. In general the gases of such gas caps contain a certain percentage of heavier hydrocarbon gases. The transition zone from gas to oil resembles that of a gas field. Within the oil bearing pay the pores contain connate water, oil, and gas dissolved in the oil. The relation of oil to the volume of dissolved gas in a certain quantity of rock is called gas/oil ratio (GOR) and may vary between one to ten or less up to one to several thousand. It depends on the temperature and the pressure within the reservoir, the type of oil and gas, and the grade of saturation. Some reservoirs are un-saturated, which means that there could be a higher GOR according to the particular reservoir conditions and the types of oil and gas. In some cases the reason for such an undersaturation seams to be an imperfect closure which is only impervious to oil but not to gas. This grade of saturation is very important for production because no gas should „cook" out of the oil within the pore space. During production, however, the formation pressure will always drop more or less rapidly. The reason for this pressure drop ought to be clear to everybody: from a more or less restricted space, – the pore space of the reservoir –, a certain volume of oil is withdrawn by pumping it out of the well. The rest of the oil will expand and show a certain pressure drop. If the formation pressure drops below the „cooking point" of the oil, gas will become free within the pore space and hamper oil production. So one of the first tasks in a newly found field is to determine the GOR of the initial production, the „cooking point" of that oil and the maximum possible GOR under reservoir conditions concerning pressure and temperature. Higher gas/oil ratios occurring during later production show that the formation pressure has dropped below the „cooking point", and that secondary methods have to be made to reactivate formation pressure by pumping water or gas or both into the reservoir.

The transition zone between oil and water is relatively thick because of the higher reciprocal surface tensions. The thickness of this zone depends on the type of oil, the formation and the temperature. In a normal sandstone with an oil of average viscosity such a transition zone will be some 3 to 5 metres thick. Fine grained sandstones have a thicker transition zone than coarse ones, and high viscosity crudes a thicker one than low viscosity oils. The transition from oil-filled pores with a certain percentage of connate water to pores with only water proceeds gradually. Bore holes drilled into such transition zones, however, will produce mainly water because the viscosity of water is always lower than that of any kind of oil.

Formation waters below or along the borders of oilfields often contain interesting quantities of bromide an iodide salts, coming from the decomposition of algae and other marine plants.

Usually the boundaries between the different pore fillings are horizontal. In some rare cases, however, where the pore fluids have a remarkable streaming potential, the boundaries are deformed following the parallelogram of forces, in this case the different gravities and the streaming potential. In extreme cases the gas cap may even be separated from the oil field by waterfilled pay.

2.5.2 The Technical Properties and the Commercial Value of Natural Gases and Crude Oil.

A well-site geologist has no possibility of making any real analysis of the gas or oil coming from an open hole test or from the first production of a newly drilled well, but he should know as early as possible what quality he has found, and about what price his company will get for it. There are some simple tests which may show him what type of petrol hydrocarbon he has found, and what the exact technical term for it is. These technical terms have a more or less direct connection with the value of the gas or crude.

Natural gases may be called sweet or sour, dry or wet. Sweet gases show a neutral reaction to a wet strip of pH-paper. They consist of hydrocarbon gases, mostly of methane, and perhaps of inert gases such as nitrogen. They will not attack steel pipes, are handy for transport and are always more valuable than sour gases.

Of course the gas itself may differ considerably in quality and composition. The highest percentage always belongs to methane, the lightest hydrocarbon gas. Heavier hydrocarbon gases like ethane, propane, butane and pentane usually show lower percentages or are even lacking altogether. Gas fields containing considerable percentages of sulphuric gases like hydrogen sulphide are called „sour", the pure ones „sweet". Sulphuric gases are very disagreeable for drilling, production and marketing. They must be stripped or washed out by chemical or physiochemical processes and converted to solid sulphur.

Other disagreeable components are carbon dioxide and nitrogen. Carbon dioxide may come from volcanic processes in the neighbourhood and is disagreeable because it attacks the steel of well pipes and diminishes the caloric value of the gas. It must be stripped off before selling, if the percentage is considerable.

Sour gases show some acidity when tested with a moist strip of pH-paper and may contain hydrocarbonic acid and / or hydrogen sulphide. Hydrogen sulphide when in relatively high percentage is the worst component possible. This gas is very poisonous for all living beings including plants, will destroy all steel and iron pipes and has to be stripped chemically before transport. On the other hand, sulphur is in overproduction all over the world, because so many gas fields produce sulphur from sour gas in enormous quantities. Sulphur gases therefore have a relatively low value. Even small amounts of hydrogen sulphide may be detected easily by their disagreeable odour, or better by a strip of lead acetate paper, which turns brown or black when treated with such a gas. Hydrocarbonic acid or rather carbon dioxide gases may occur near volcanic regions, or over coal fields. They may be stripped off more easily than sulphur gases, but anyhow they may attack steel pipes if the gas is not completely dry, and any remarkable amount of that gas will reduce the value of the production. Carbon dioxide will produce a white precipitate when it is passed trough a clear solution of quick lime, i.e. lime water.

Dry gases are those which do not contain condensate, while wet gases contain considerable amounts of condensate gases like propane, butane, pentane or even hexane. These heavier hydro-carbon gases turn to the liquid state at low temperature and at high pressure, and may form liquid drops in pipelines in winter or cause difficulties in underground storage facilities. On the other hand these condensate hydrocarbons will bring good prices only if an organisation is near which will fill them into bottles and sell them, or if there is some chemical industry in the vicinity which will buy that part of the production. In general they are troublesome and reduce the value of the gas production.

Best prices are paid for a dry, sweet gas, like methane. Of course a well-site geologist will not have any facilities for testing the percentage of methane, but he may see the general composition when the gas is flared off during an open hole test or the first production test. Dry methane will burn with a bluish flame with yellow borders. Wet gases, however will burn with a red flame with a dark, smoking border. Of course a good gas logging unit with a gas chromotographer will produce more and better details, but with the usual hot wire detectors it is difficult to establish a sure relationship.

Many gas fields contain remarkable amounts of nitrogen. There are some regions in North America, Hungary and Middle Europe where the percentage of nitrogen increases in one direction gradually until some extreme fields contain only nitrogen, while fields in the opposite direction may contain only pure methante. The reasons for such regional distribution are unknown, but of course nitrogen cannot be sold, and will reduce the price of the gas. Percentages of more than 45 of 50 % nitrogen are the outer limit where such gases may be sold for heating and the production of energy, and their heating value is much lower of course than that of pure or nearly pure methane. A wellsite geologist usually has no possibility of stating what percentage of nitrogen he has found. That may be found out with a gas chromatographer or some other suitable gas logging unit. In general the only test is to state if the gas will burn or not, and if it burns it will fetch some money too.

Sometimes there is a small amount of helium, crypton or argon within the inert part of the gas. Helium is especially valuable and will increase the price considerably, even if the percentage is relatively low.

In principle all newly found gasbearing structures have a commercial handicap: Gas can be sold only when the size of the structure and its gas contents are known, because the diameter of the pipe lines, the size of the installations and the quantity to be sold per day or per year must be known before a selling contract with some industry or township can be signed. That means that a big amount of money, paid for drillings and installations will not bring in any income or interest for years until the real production begins.

Crude oils usually are classified by their gravity, their pour point, their flame point and their sulphur contents. The gravity is one of the most valuable tests, and it has a direct relation to the price. It may be tested easily by an areometer. In the United States the gravity of

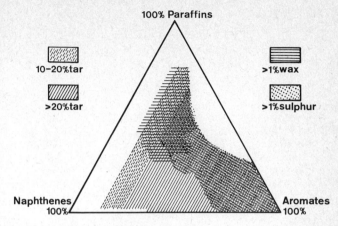

Fig. 24 Paraffinic, naphthenic, and aromatic types of crude and their affinities to wax, tar and sulphur. (After Luther.)

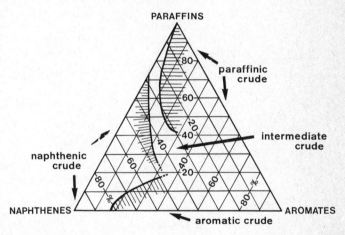

Fig. 25 The nomenclature of some different types of crude oil. (After Luther.)

crudes is measured in API-grades, in Europe usually in centigrades. Low gravity crudes may show values as low as 6,6 or 7 centigrades at a temperature of 18 centigrades. They contain a high amount of low-boiling components and will give a high percentage of gasoline by a simple refining process. So they are sold at a higher price, especially if they are free of sulphur. Average crude oils vary around 8 centigrades of gravity. Some crudes even show more than 9.5 or even 1.5. That would mean that they are heavier than water, but within the structure, under higher temperature and with some dissolved gas they are always lighter than water.

The API-grades vary between 1 and 100, comprising all types of crude oils. Low gravities correspond to high API-grades, and vice versa.

High gravity oils will bring a lower amount of gasoline when they are refined, and special cracking processes have to be used to increase the amount of better-paid refining products. High gravity oils usually contain a fairly high amount of naphthene compounds which are troublesome for the refining process, and in general they fetch a much lower price than low gravity or medium gravity oils. Sometimes even medium gravity oils are sold better than low gravity crude, especially when the price for heating oil is high, or when the refining plant wants to get a high percentage of lubricating oil. These variations, however, depend on seasonal and regional fluctuations.

The pour point of crude oils is most important for treating, transportation, handling and refining. The pour point means the temperature where a liquid crude becomes a stiff smear which will block all tanks, pipelines, treaters and installations. Handy oils have a pour point of less than 5 or ten centigrades temperature, what means that they behave like water or similar liquids. High pour points, however, mean that the oil has to be heated during pipeline transport within the field, that heater treaters have to be used for the separation of gas and water, and that pipeline transport may become very difficult in winter or even on cold nights. There are crudes with pour points of more than 20° centigrades which will become stiff even on cold summer days in a temperate climate. The reason for such behaviour is mostly paraffin compounds which form solid needles and increase the viscosity of the crude until it becomes unable to flow or drop. Such high viscosity crudes cause a lot of trouble within the oil field itself and during transport. They cause higher expenses for production, and are sold at a lower price.

The flame point of crudes may vary between about 80° and less than zero centigrades. Low flame points show that the crudes contain high amounts of low-boiling hydrocarbons such as benzene, and make such types of crude more valuable than high flame point types. The detection of the flame point is made in special facilities, and there are no simple tests available which could be made in a small well-site laboratory.

The sulphur content of crudes may go up to more than ten percent; usually with high gravity, high viscosity and high flame point types of oil. Such high sulphur contents usually occur near salt domes, or in series containing gypsum or anhydrite. The sulphur may occur in a solid state, it may be dissolved within the oil, or in the worst case it may occur as a polysulphuric compound like mercaptane or simular compounds. Solid and dissolved sulphur may be stripped from the oil by relatively simple processes during refining. Mercaptane and similar compounds, however, become volatile during boiling and move into every fraction of the refining products. They must be removed by chemical processes which are relatively expensive, and are one of the worst factors of air pollution by heating oil. Most Arabian crudes contain considerable amounts of sulphur and sulphur compounds.

Every tenth percent of sulphur will lead to a lower price of the crude, therefore, and this rule will become still more important when the laws about air pollution forbid higher contents of sulphur dioxide in the exhaust of chimneys.

2.5.3 *Primer Structural Types of Reservoirs*

Text books an petroleum geology show different methods and possibilites of fitting the innumerable types of oil- and gasfields into more or less sophisticated and complicated systems. Many of these systems resemble the one which is shown in this booklet, others show a certain mixture of structural and causal points of view. The system listed below shows a clear differnce between the type of structure itself and the causes which led to its creation. It has the advantage of clear and logical separation and easier understanding.

The primer types are

anticline type reservoirs,
fault type reservoirs,
disconformity type reservoirs and
stratigraphic or facies type reservoirs.

The causal types are

reservoirs caused by salt domes
reservoirs caused by reefs and bioherms
reservoirs caused by buried hills
reservoirs caused by movements of geosynclinal
or graben structures,
and others.

Anticline Type Reservoirs

The simplest and most instructive type of reservoir is that of an anticline, where the pay is bent downwards on all sides, forming roundish or elliptic contour lines. Petroleum hydrocarbons may

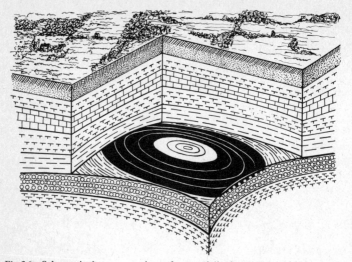

Fig.26 Schematical cutaway view of an anticlinal structure with a gas cap (light), oil (black) and formation water (small circles).

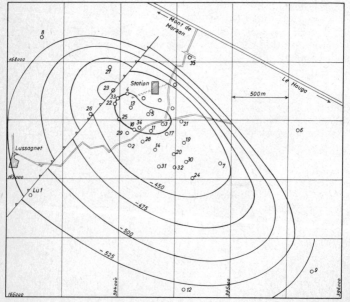

Fig. 27 The anticline of Lussagnet (France), which is used for the aquifer storage of gas. The structure is nearly perfect, except for a small fault.

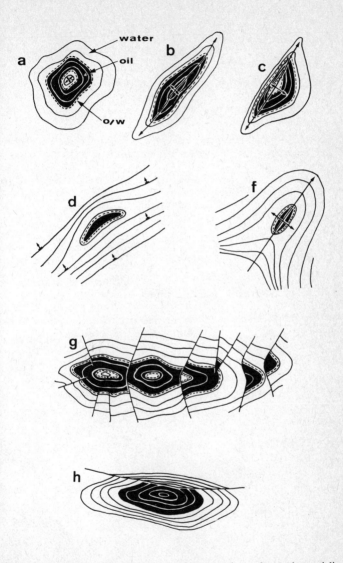

Fig. 28 Different types of anticlines. a) a normal, nearly regular anticline. b) a long, elliptical anticline. c) an anticline with unequal dip on both sides of its axis. d) a so called terrace. f) a so called balcony. g) an elliptical anticline, complicated by some faults. h) an elliptical anticline with a fult running nearly parallel to its axis (After Flandrin).

migrate into such a structure from all sides, accumulating in the top of the structure where the impervious roof formation brings the migration to an end. Usually another impervious layer lies directly under the pay so that gas, oil, and water are in contact only along the flanks of the structure.

Such anticlines may be formed by tectonic movements like a gentle folding, by the movement of salt domes, over buried hills, over buried reefs, and over sediments with a smaller rate of settling, e.g. over sand bodies in shaly surroundings.

The structure of such anticlines may be complicated by faults, changing porosity of the pay, by special bending and folding, and by other factors. Anticlines in general can quite easily be detected by geologic and geophysical methods, and such fields are relatively simple to operate and to develop. There are giant fields of anticline type with thousands of producing wells, and small ones with only a few bore holes.

Fault Type Reservoirs

Other reservoirs are formed when a fault plane interrupts the direction of migration. Most common are reservoirs along so-

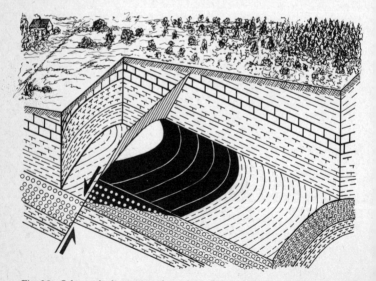

Fig. 29 Schematical cutaway view of an oil trap before a reverse fault, with a gas cap, oil, and water. The trap contains oil though the pay formation of the higher block is in direct connection with that of the lower block.

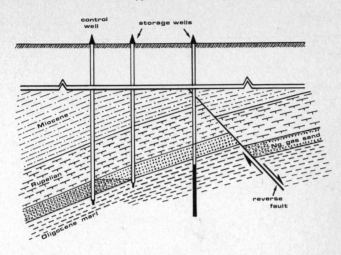

Fig. 30 Cross section across such a structure with a reverse fault. This structure above the salt dome of Reitbrook (Germany) is used for the aquifer storage of natural gas. (After Behrmann.)

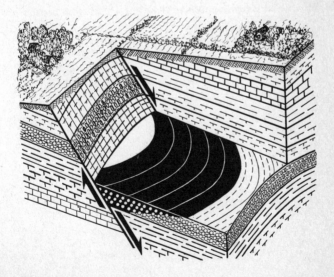

Fig. 31 Trap before a "normal" downthrust fault. The trap would be empty, if the pay formation of the lower block would be in contact with that of the upper block.

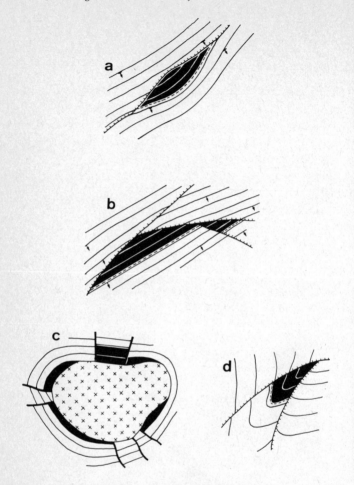

Fig. 32 Traps before, behind and between faults. a) A trap before a reverse fault. b) Traps before and between some reverse faults. c) Radial faults around a salt dome. d) A trap between two ramifying faults. (After Flandrin.)

called reverse faults which on their part are common in basin and graben structures, where they mostly are parallel to the axis of the structure. The reverse faults are formed by rotation of sediment blocks when a basin or graben sinks down and is filled by more and more sediment, while one border or both borders remain relatively stable.

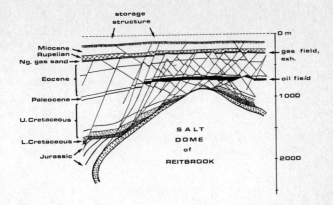

Fig. 33 The oil field of Reitbrook (Germany). The pay formation consists of limestones belonging to the Upper Cretaceous. Movements of the salt dome have created a very complicated pattern of smaller and bigger faults.
(After Behrmann.)

Such so-called geosynclines are found at the foot of nearly all major mountains. Reverse faults in such basins may be hundreds of kilometres long and carry oil fields lined up like pearls on a necklace.

The technique of accumulation is fairly simple. The hydrocarbons coming from the centre of the basin and migrating towards the higher borders enter a fault block in its lower part and are blocked in the higher part, where the porous and permeable pay is thrust by the fault against impervious layers. The closure in such a type of field is the height between the top of the structure and the uppermost point where the pay of the upturned blocks is in full contact with that of the downturned part of the next block. So there will be some accumulation even if the pay is connected across the fault plane within the top of the structure, the most striking difference of such reservoirs from those on normal faults.

Normal faults are widespread among all kind of tectonical structures, but usually they are shorter and their direction follows other laws. Most common are such normal faults in the immediate neighbourhood of saltdomes and similar round structures, where they mostly show a radial tendency.

Along such faults an accumulation of hydrocarbons will occur only if the pay of the upper block is completely separated from that of the lower block by the fault plane in the highest part of the structure. Otherwise there would be no interruption of migration by the fault plane, and the oil or gas would proceed to the next higher trap.

Reservoirs along overthrusts of every type obey similar conditions. Most common are such overthrusts along the flanks of salt domes, between the mass of older salt and anhydrite and the younger surrounding sediments. If such sediments are turned up and thrust against the anhydrite mantle of the salt dome, hydrocarbons may be trapped along the plane of the overthrust and form a reservoir.

Rather different are structures below the thrust planes of big nappes like those in the Canadian foothills. There the upper and older sheet of sediments covers younger and partially porous sediments, trapping the oil and gas that is ascending in them to higher zones of lower pressure. Such reservoirs are more like those which are formed by transgressions.

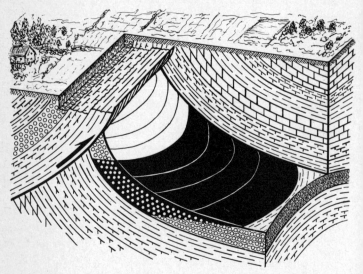

Fig. 34 Trap below an overthrust fault. The shales of the upper block seal the pay formation of the lower block.

Disconformity Type Reservoirs

Two special types of reservoir are found on or below disconformities, those above and those below the plane of transgression.

Most common are fields below a transgression, the others, called overlap fields, are fairly rare.

Disconformities are created by relative movements of the level of the sea. If the level sinks down, the shore line will move outwards. During the time of regression the former horizontal layers may become folded and tilted by tectonic movements. If the sea level increases in height again, the shoreline will move inland, levelling the hills and mountains that have been formed in the meantime and covering them with new sediments, which have more or less horizontal bedding planes. The angle between the bedding planes of the older and the younger generation of sediments is called disconformity, and the boundary between them transgression plane.

Porous layers in the older block of sediments will be cut open by such a transgression, opening possible paths of migration for some time to the open sea. As soon as new sediments cover such an outcrop, hydrocarbons may accumulate below the transgression plane, thus forming a reservoir below a transgression. Such fields are fairly common in basins and grabens whose history covers more than one phase of movement. If the angle of disconformity is small, it needs very exact and sophisticated seismic work to find such fields in deeper horizons.

Overlap fields may develop when a disconformity is tilted or bent, and when new layers cover the sinking interior of such a basin or graben. Towards the borders each new layer overlaps the older one, and if there is a porous layer in between, it may become an overlap field.

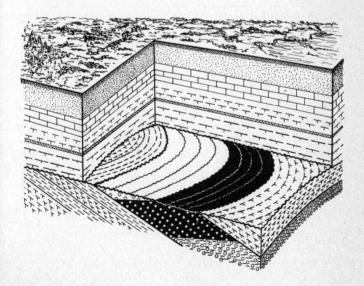

Fig. 35 Trap below a disconformity. The lower and older block with its pay formation has been tilted, eroded and covered by younger sediments of the upper block.

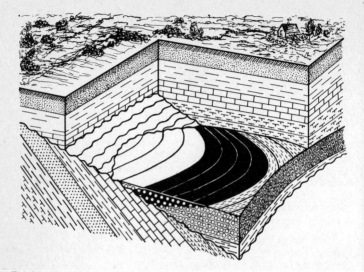

Fig. 36 Trap above a disconformity, or overlap field. Both the older block an the plane of the disconformity have been tilted. The younger sediments overlap one another. The pay formation is sealed by overlying and over-lapping shales.

Stratigraphic or Facies Type Reservoirs

This type of reservoir may develop whenever a porous sediment is bordered by an impervious layer between the same bedding planes or vice versa. If such a boundary of facies blocks the direction of migration towards higher zones of lower pressure, petroleum hydrocarbons may be blocked in what is commonly called a facies trap. Comparisons of such fields show that two types of such traps may be distinguished, those caused by sedimentation and those caused by diagenesis, i.e. by alteration of rock by internal processes.

Facies type reservoirs caused by initial sedimentation are mostly of the „shaling-out" type. A porous layer, sand, sandstone, limestone or dolomite changes over to impervious shale or another tight facies. If this boundary is situated in the right position, hydrocarbons may accumulate below it and form a „shaling-out field".

Other structures trapping oil or gas may be formed when porous layers are rendered impervious by diagenetic processes, e.g. by deposition of calcite, quartz, gypsum, iron minerals, or even asphalt or paraffin. Solutions of the first-mentioned minerals always circulate with the formation water within the pore space and may block it completely under certain conditions. Boundaries of impervious sediment may be created by the precipitation of such solutions, giving rise to the accumulation of petroleum hydrocarbons. Other types, the so-called asphalt props, are restricted to warm and arid climates only. Oil within the pore space of outcrops of porous layers into the open air may be thickened and turned into asphalt by evaporation, oxidation and polymerisation. Such structures where the oil content is preserved by an asphalt prop are still fairly common in the Near East, and they formed the famous tar pits in Mesopotamia at the time of Nebuchadnezzar and the tower of Babel.

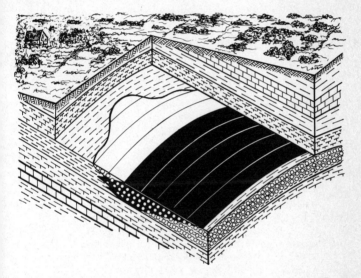

Fig. 37 Oil trap of the shaling out type. The pay formation is porous and permeable only in its lower part, the upper part is shaly and impervious.

Reservoirs Containing Several Trap-Forming Elements

Many oil- and gas-fields contain more than one of the above-mentioned elements, and often the trap itself is formed by a combination of such elements. In most cases only one of the above-mentioned types forms the real trap, while the other structural elements only complicate the contour lines and the shape of the field. There are other fields, however, which would not form any reservoir if there were not a combination of two or more of such elements. So there are for example fields which are limited on one side by a facies boundary, on the opposite side by a fault, along the slope by a disconformity and which are only open on to the water margin. Other fields are even closed on all sides, which may create severe difficulties for the production of the contents unless suitable secondary methods are used.

In general there are so many possibilities of similar and still more complicated combinations that it would be too extensive to mention only a part of them. For the exploration it is much more important to know which types of reservoir are to be expected in a certain basin or major structure and how they may be found by geologic and geophysical methods.

2.5.4   The Most Common Types of Reservoir in Geosynclines and Graben Structures

The significant structural elements within geosynclines and graben structures are reverse faults, running parallel to the axis of the basin and forming excellent traps for the oil and gas coming from the sediments of the basins themselves or from older sediments in the base. Other elements typical of such regions are shaling-out structures. If the development of the basin covers more than one tectonical phase, reservoirs could occur below a disconformity and similar structure. Reservoirs along normal faults may be anywhere and are not restricted to particular regions or basins.

The Saxonical basin in northern Germany with mesozoic sediments shows the effect of a tectonical phase within the Lower Cretaceous. Quite a series of reservoirs is situated along the pinchout of the older mesozoic sediments below that disconformity. Other types of reservoir are rare in this region and restricted to the salt domes. Basins over very stable and solid regions may look very different, showing less detail and nearly no faults and internal structures. Others may be much more complicated, especially if the basins begin to become „ripe" in the gelogic meaning of the word and show the beginning of folding and narrowing. Then anticlines of every kind and shape will become most frequent and striking.

2.5.5   Types of Reservoir Caused by Salt Domes

Salt Domes occur in many major basins and are the most polymorphic structures as regards their shape, size, age, „personal" history of development and their effect on their geologic neighbourhood.

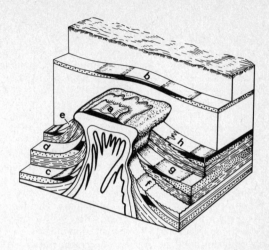

Fig. 38   Oil traps above and around a salt dome. a) Caprock field of anticlinal type within the top of the salt dome. b) Field of anticlinal type in layers above the dome. c) Fault type field at the flank of the salt dome. d) Fault type field below the overhanging top. e) Fault type field between two radial faults. f) Field below a disconformity, which has been caused by movements of the salt. g) Overlap field above such a disconformity. h) Shaling out fields on one side of the salt dome.

Their shape may vary from a slightly inflated cushion, a hill with soft flanks, a steep dome with overhanging cap, to a separate salt mass that has lost any connection with its former formation. The form of such diapirs may be round in cross section, irregular or extended like a volcanic dike.

Their size may vary from a few cubic kilometres to enourmous structures with a length of several hundred kilometres and a width of several kilometres. Their age may vary from early Paleozoic to youngest Tertiary, and their „personal" development may cover only one phase of movement as well as a series of different phases of movement and of standstill, somewhat resembling the tectonical phases. Their effect on their geologic neighbourhood is as varied as their abovementioned properties, and nearly as varied are the types of hydrocarbon reservoir found over, on and in the vicinity of salt domes.

The simplest way of trapping oil or gas is within the so-called cap rock in the uppermost part of a salt dome. This cap rock is a kind of breccia consisting of fragments of anhydrite, gypsum and limestone. As to the origin, this breccia represents the insoluble residues of the uppermost part of the saltdome after being in contact with formation water. Empty space between the single fragments and within the mostly porous limestones may be filled with oil and gas coming from sediments of the basin around the salt dome. Of course the cap of the salt dome must be covered by a layer of impervious sediments. Other layers above the salt dome may be bent by salt movements to form an anticline, so that there are numerous possibilities of forming reservoirs in sediments above a salt dome. Very common are reservoirs along the flanks of a dome, which mostly belong to overthrust type fields. Their crucial trap is the overthrust of the salt body over the surrounding sediments. Along the flanks of a dome the sediments often have been bent up and even overturned, forming nearly ideal trapping possibilities in porous sediments. Other traps may occur along the radial faults directed from the salt dome into the surrounding sediments. They are mostly of normal fault type. Facies type reservoirs may be formed directly by movements of the dome during sedimentation in its surroundings, or later on by precipitation of gypsum or sulphur in porous layers. Quite common on the other hand are disconformities in the surroundings of a salt dome, caused by salt movements during sedimentation. Traps for petroleum hydrocarbons may occur above or below such disconformities. Even at some distance from a salt dome there may be reservoirs of different types caused by tilting and bending of sediments by salt movement at greater distances.

2.5.6   Reservoirs Caused by Reefs

Reefs or bioherms, built up by various animals and even plants, occur in all formations from the Precambrian up to recent times and are very important for petroleum production, especially in the Near East. Limestones and dolomites of such fossil reefs and bioherms in average have an enormous porosity and even a very high permeability which are the very reasons for the amazing production rates of such wells and fields. The high porosity of such a reef complex is for the most part of primer and initial origin, coming from the cavities within the shells of the reef-building organisms and between the shells or tests of different reef-builders. From recent reefs we know that such complexes contain a lot of caverns and gorges which will be filled with loose fragments before they become fossil. Other, secondary porosity and permeability may be created by solution of lime or transformation of lime-, calcium carbonate-, into dolomite, the carbonate of calcium and magnesium. Even tight limestones may become porous by this transformation, because the newlyformed mineral takes up about 10 % less space than the former one.

Reefs, therefore, are primer reservoir structures, if they are situated in an oil-bearing basin, and if they are covered by impervious strata. Then oil and gas may accumulate in the top of the reef, resembling the anticlinal type reservoirs. In Canada we know of series of Devonian reefs; the reefs of the „Golden Lane" in the Gulf of Mexico are of cretaceous age, and the Asmari Limestone in the Near East is Cretaceous too. Algal bioherms of tertiary age are a very important pay in the Molasse basin in eastern Bavaria.

But not only the reef complex itself is important for the formation of reservoirs. Sediments above and around such a reef may be affected in their sedimentation and in their structure in many ways, causing a lot of different possibilities of forming traps. Sediments above such a reef are mostly deformed by the settling of the soft sediments around the reef, forming anticlinal type reservoirs.

Around the reef there is some probability of disconformities with corresponding creation of traps, and layers dipping away from the reefs may form facies type structures.

2.5.7   Reservoirs Caused by Buried Hills

Buried hills are common below disconformities which cover older and consolidated formations. They will, for example, develop when a transgression proceeds over old and solid rock without levelling it completely. The new and younger layers above the disconformity will cover and bury such hills, so that oil accumulations in the top of the hill are of disconformity type (reservoir below a disconformity). If such a hill consists of crystalline and not of porous rock, there may be accumulations in granite wash on top of or along the flanks of the hill, mostly of anticline or pinchout type.

Reservoirs above a transgression may develop especially on the sides of a hill where overlapping younger layers touch the plane of the disconformity. On the other hand the shrinkage of the younger sediments may cause reservoirs of anticlinal type in sediments above the hill and along various types of faults which are caused by different rates of shrinkage.

So nearly all primer types of reservoir may be found on, above and around buried hills. The only type that really should not occur is traps along overthrust faults.

Buried hills may be found by gravimetric surveys and by seismic investigation. They are common reservoir types such as along the disconformity between older and younger Miocene in the basin of the river Po in Italy, where they bear the most important gas fields of that country.

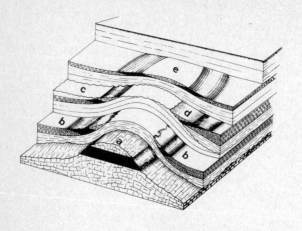

Fig. 39   Traps above and around a buried hill or a reef. a) Trap of anticlinal type within porous parts of the top of the hill or reef. b) Traps of the shaling out type along the flanks of the buried hill or reef. c) Trap below a disconformity. d) Overlap structure above a disconformity. e) Anticlinal type structure in sediments overlying the buried hill or reef.

3   The Structure of an Average Oil Company

Oil companies may be big or small, they may have a long tradition or change their name every few years for formal reasons. They may belong to the government of some country and work without regard for money and the rate of return, or they may belong to financial groups or private persons and try to work at the highest possible profit.

In principle such a company could sonsist of a director and a small managing staff only. Then all special work would be done by contractors, from exploration to the marketing of crude and gas. Other companies perform nearly all the work themselves, topographical and geological mapping, geophysical surveying, drilling, logging, well completion, production, transport, refining and marketing. Other companies, such as the Italien Ente Nazionale Idrocarburi, even produce drilling rings in workshops of their own, own big petrochemical plants and train their collaborators in schools of their own. An average company will usually comprise a top management staff, a financial department, a legal department, a department for exploration, a department for exploitation, a marketing department and a department for technical and general services, workshops and so on.

Geological work is done in the exploration and exploitation department, and these departments work in close cooperation of geologists, geophysicists, drilling engineers and production engineers. Producing fields and drilling rigs usually are spread over various regions and countries, and smaller and bigger teams of these departments have to work there in temporary offices, camps or even long-term regional offices whereever they are needed.

The exploration department usually comprises a geologic, a geophysical, a drilling branch and a small branch for production engineering. All these branches in the central office have a corresponding staff in all or most regional offices, and on all drilling sites or producing fields whereever they are necessary. The smallest working team may be a jeep with a topographer, a geophysicist and a helper making a gravity survey, or a driller with his rig, his roughnecks, a wellsite geologist and some samplers. At the very moment, however, when such well strikes some interesting pay, a lot of problems about logging, testing, well completion and other things show up which need the help of specialists from the regional or central office by radio, telephone or on the site. All these problems have to be resolved by teamwork, because even the smallest problem may have a technical and a geological, a local and a regional side. A geologist who works in the exploration branch has to be first and foremost a stratigrapher with a good understanding of sedimentology, of geophysical methods, and he should understand a lot about drilling techniques in order to be able to work well together with his colleagues.

The so calles *samplers* mostly do routine work on the drill site, washing and sieving samples, making statistics about the sedimentology and the micropaleontology of the samples, and most of the time they even run the gas detector, which registers or records drops of oil and bubbles of gas contained in the drilling mud. Sometimes they even check the drilling mud for its chemical and technical properties, a technique that has become a real science since deep drillings go down to more than 6,000 metres even in solid rock and 10,000 metres in young, hardly consolidated layers. The temperature, pressure and the expense of such drillings made it necessary to invent other types of mud than only clay and water.

The *exploitation department* comprises a branch for geology, production engineering, transport, marketing and usually one for technical services, workshops and similar things. Their smallest unit usually is an oil or gas field, or a newly productive wildcat with only a few men. Developing and producing a field can be done by team work only, like in the exploration department. The geologist has to map the pay and its properties, with the aid of evaluating borehole logs and labour analysis. The spacing of the wells needs the cooperation of a production engineer; and all drilling, testing, completion and production is done in close cooperation between engineers and geologists. Of course the type of work done by a geologist in such a producing field is quite different from that of a

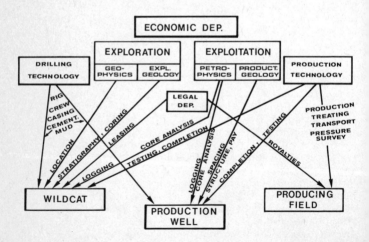

Fig.40 Organisation of a typical oil company, and the main principles of cooperation between its different branches.

wellsite geologist who runs an exploratory wildcat. He should be half a sedimentologist, a good logging specialist, he should have good contact with the laboratories and their facilities and he should have a good understanding of production techniques, petrophysics, and the physics of fluids and gases in porous media.

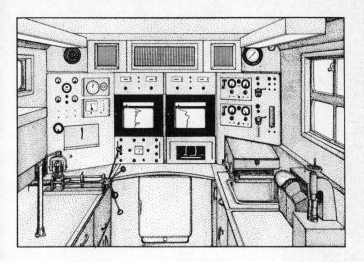

Fig. 41 The interior of a wellsite gas lab. The lab contains mud gas analysers and recorders and facilities to treat and to analyse cuttings and cores.

3.1 Geological Laboratories and Specialists

At every drilling rig there is a small portable laboratory which is carried from drillsite to drillsite on a trailer or on a lorry. These small laboratories contain everything necessary for routine work, a washing tray with sieves for washing samples, hydrogen peroxide for the disintegration of claystones and marls for micropaleontology, binocular lenses and microscopes, an UV lamp for recording oil traces in the drilling mud, a gas detector for the gas survey of the drilling mud, and all facilities to store papers, samples and cores of that well. In far-off locations such laboratories even contain the necessary equipment for porosity and permeability analysis of cores with a drilling machine, a diamond saw, a porosimeter and a permeameter, and some equipment to make thin slices of rock. Core analysis in general needs a lot of machines, equipment and trained specialists, so it is done mostly in central laboratories with the necessary facilities, if transport is not too expensive. Other central laboratories contain specialists for micropaleontology, which control and improve the work done by the samplers, drillsite geologists and mapping crews. Other laboratories work on sedimentology, bore-hole logging, drilling mud control, oil and gas analysis and such things which can be done by experienced men only who are not as much under the pressure of time as for example a drill site geologist. All these laboratories and specialists may work as well merely scientifically, in mere routine, or as so called trouble shooters who have to solve special problems which hamper the work of some colleague in an outside or regional laboratory. Some big companies even have specialists working on themes of fundamental interest, such as the genesis of petroleum, which are without any direct importance for production or exploration.

4 The Principal Laws of Petroleum Exploration

In principle, petroleum exploration is a complicated mixture of geologic, geophysical, chemical and technical methods. There are three or even four fundamental laws, however, which enable geologists to say if a certain region, a certain sedimentary basin or even an entire country contains petroleum hydrocarbons or not, if it is worth paying several million dollars for leasing, and if it is worth an expensive exploration campaign.

The fundamental laws are:

1. The region in question has to contain a sedimentary basin with enough marine, brackish or even limnic claystones and marls which can have served as mother rock. Terrestrial, volcanic and crystalline rock are of course not good prospects.

2. The sedimentary series should contain some porous layers which may serve as pay, e.g. sands or sandstones, porous or vuggy limestone or dolomite, and similar kinds of rock with apparent porosity and enough permeability for the migration of petroleum and the formation of reservoirs.

3. The region in question should contain some structures which may lead to the formation of reservoirs, e.g. faults, disconformities, reefs, buried hills, salt domes or similar geologic units. There should be a certain lapse of time since the last major tectonical movement of this region, so that possible reservoirs can have been filled by migration of oil from older accumulations.

4. The sediments of the region in question should be free of metamorphism, even to a slight extent, which would distillate petroleum out of its reservoirs, modify and even destroy porosity and permeability, and block the normal ways of migration.

In earlier times of oil prospection the principal law was that such a region had to contain oil seeps, oil and gas seepages, traces of oil and asphalt in drillings, mines or tunnels, and similar obvious evidence of the presence of petroleum hydrocarbons. This so-called law, however, has proved to be only partially true, especially if used for deep basins.

There are many regions on the world without any seepages, oil seeps, oil traces, gas seeps and similar occurences which on the other hand contain big oil and gasfields in the underground. On the contrary such seeps on the surface show that the reservoirs from which the oil or gas comes are not tight, that their contents are running out or have already run out. Such regions may be devoid of any profitable reservoirs, but cause a lot of expense for continous wildcat drilling boostered up again and again by new traces, by oil bleeding but waterbearing sandstones, and similar references to reservoirs. The only way to success in petroleum exploration is continuous, constant and exact work with all possible geologic, geophysical and geochemical methods until the principles of reservoir formation for the region in question have been found out and wildcats can be drilled with regular success.

The average rate of success of wildcats is about one in ten, that means that nine out of ten wildcats remain „dry" and have to be abandoned. Wells with small production often become more troublesome than dry ones, because most regional and national laws force the company to produce them without regard to the profit. On the other hand profit or not profit depends on fluctuations of price and marketing conditions. A well in a desert region with a production of 20 tons daily will usually remain non-profitable. A field of some hundred of such wells however may bring some profit if the type of crude is good or averrage, and if transportation expenses are not too high.

4.1   Some Words about Methods of Petroleum Exploration

Petroleum exploration is teamwork of geologists and geophysicists, in some cases even of geochemists. A lot of different methods have been abandoned and have been used again after some improvements which rendered them more useful or more practical. Thus geomagnetics have been used for some time to outline saltdomes, buried hills and other structures hidden under a cover of sediments. Geomagnetics, however, need a lot of calculation and comparison with stationary magnetic stations if their results are to be reliable. Such surveys on foot therefore need a considerable lot of time before a drilling location can be fixed. So geomagnetics have soon been replaced by gravimetric and seismic surveys. When airborne magnetics were invented for military purposes, however, and proved to be useful for geological exploration too, they soon became a useful method for petroleum exploration again. It cannot be the purpose of this booklet to mention and outline all these exploration methods, to discuss their theory and their usefulness. Readers interested in further detail should read a special textbook on this subject.

The following pages only contain some geological aspects of the most usual methods and some remarks which could be useful for younger colleagues.

4.1.1   The First Steps of Petroleum Exploration

Petroleum exploration always begins with an exact topographic and geologic mapping of the region in interest. Of course this work has to be done in reasonable time, and with only restricted interest in geologic surface mapping. In regions, where all older and interesting structures are covered by young, terrestric layers, such as glacial deposits in Northern Europe, the classification of these deposits will be of minor interest for a petroleum company.

The first survey of unknown regions will always be done by airfotogrammetry and fotogeology. These methods allow the exploration staff or special contractors to map a region very quickly both topographically and geologically. The first geologic maps, however, have to be dated and controlled by small groups of geologists on foot, in jeeps or in helicopters. In some regions, oil bearing formations pinch out at the surface under only a slight cover of younger sediments. In such countries anticlines, faults and other structures may be mapped directly from air fotos, and the subsurface geology seems to be open to the human eye. Of course even such obvious structures need some more geologic work before a drilling location can be fixed, because the angle of the dip of the axis, and other special features must be mapped and cleared before the real subsurface can be reconstructed.

4.1.2   Big Scale Exploration by Airborne Magnetics

In earlier times, magnetic methods have been used in petroleum exploration in some rare cases only. After the First World War, however, airborne magnetics became an important tool for quick big scale mapping of entire basins and of magnetic anomalies hidden in their underground. Meanwhile these methods have been modified and improved by newly developed caesium magnetometers and by proton-processiometers. One of the greatest difficulties of all geomagnetic measurements are the permanent and irregular variations of the earth's magnetic field which may cover and overlap anomalies caused by geologic structures. Magnetic crews working on foot therefore always use two stations simultaneously, a stationary and a moving one. This system would not work with airborne magnetics because the distance between a stationary magnetometer and a plane would surely be too great for exact comparison. The plane therefore has to fly according to a complicated pattern, either a spiral line or in parallel lines. With these patterns the plane will cross the base line where it began its flight and where it will end it, so that any variations during the time of flight will appear on the diagram.

Another difficulty is the location of the plane during the time of flight. Optical means usually are not adequate for greater distances, so electric means like LORAC and similar methods are used.

The magnetometer hanging on a rope some metres behind the plane – in order to avoid the plane's magnetic field during the logging flight – is called „bird" and transmits magnetic anomalies to the plane in the form of a continuous diagram, which is stored in a small computer. The position is checked automatically and stored likewise in the computer. After the flight all data are retrieved from the computer and may be controlled and printed automatically.

Airborne magnetic maps show a lot of general trends in a sedimentary basin, and usually a lot of details too. If the sedimentary basin is situated on a cristalline basement, the depth of the basin, the axis and other general trends can be outlined. Buried hills of cristalline rock, saltdomes and similar units will show up as positive or as negative anomalies.

4.1.3   Surface Gas Logging

In 1923, two German geophysicists showed in a short paper that the earth above oil and gas bearing structures may show gas anomalies. Theoretical investigations into the reasons for such anomalies found out that these structures are impervious to oil and gas in their normal state, but not to water in very small quantities and in geologic times. So some minute quantities of water may arrive at the surface, and if they contain dissolved gas, the gas will become free when the water gets vaporised at the surface and may be investigated by sensitive gas detectors.

Of course there are a lot of errors possible if such a method is applied, especially if it is used as the only method of exploration.

Methane over swamps, or methane coming from layers of peat, lignite or other coal seams may cause anomalies too. They may be distinguished from those caused by petroleum reservoirs by the different composition of the gas. Gas anomalies over petroleum structures usually contain noticeable amounts of heavier hydrocarbon gases which are rare or absent in anomalies caused by caustobioliths. Any gas detectors used for such surveys must be fitted out to show such differences, at least the ratio methane/(ethane or methane) propane. Another difficulty may be caused by porous layers in the underground which contain only water with dissolved gas, but which do not contain any reservoirs. All real reservoirs below such „gas-bearing" horizons will disappear under the constant gas level of gas-saturated, waterbearing horizons. Anyhow there seem to be some regions in the world where this method of „gas sniffing" has led to some success or may lead to success. In regions with weak tec-tonics, where oil bearing anticlines nearly come up as far as the surface and are hidden by some hundred metres of younger deposits only, this method may be really valuable.

It seems to be used all over the East Block and in all countries which depend on the U.S.S.R. (Union of Soviet Socialist Republics) but there are no reports on a real success in countries outside the East Block.

4.1.4 Big Scale Exploration by Gravimetric Surveys

Gravimetres are very handy and inexpensive tools to measure larger regions in a relatively short time and to locate gravimetric anomalies caused by salt domes, by buried hills consisting of dense material, and to map outcrops of tight and heavy sediments below a flat cover of young horizons.

The principle is quite simple, and even the apparatus itself is handy ad relatively cheap. A gravimetric crew consists only of one geophysicist, a topographer and one helper. The gravimeter may be installed on board a jeep or a car, and the biggest difficulties are to get a first class map and the exact location of every measurement, because all topographic factors may affect the results, especially in mountains and hilly regions. In earlier times all these necessary corrections had to be calculated by hand and took much more time than the work in the field. Nowadays all results from the field-work are fed into a computer and may even be printed out on maps automatically. Such gravimetric surveys, after making all necessary corrections, usually produce a map of the so called Bouguer-anomalies, which in short are those anomalies caused by smaller geologic structures in the underground, and which could be interesting for the location of petroleum reservoirs. Salt domes appear as negative anomalies, because rock salt is of a lower density than all consolidated sediments. Buried hills of high density crystalline rock appear as positive anomalies, and even outcropping conglomerates will show higher values than surrounding claystones and marls.

In general, gravimetry is a handy and useful means of getting a first impression of a sedimentary basin and to outline its structures and possibilities. Before a reservoir can be found, however, other and more precise methods must be applied.

4.1.5 Big Scale Exploration by Refraction Seismics

This method war invented in 1924 by *L. MINTROP* and proved to be a very useful tool for big-scale and even small-scale exploration. Since the Second World War it has mostly been replaced by re-flexion-seismics, but it is still applied for big-scale surveys in unknown regions and for some special purposes.

One of the advantages of this method is the long distance covered by every shot, the biggest disadvantage is the impossibility of interpreting refraction diagrams in regions with complicated geologic structures.

The principle of this technique is fairly easy to understand, though its theory caused a lot of headache to physicists until it could be cleared up not long ago. The necessary crew and equipment comprise one or mor geophysicists with some helpers, a lot of special geophones, a registration car with seismic recorders, a driller with some helpers and a drilling rig for smaller depths, a topographer and last but not least a connection between the shot-hole and the registration car by radio or a temporary telephone line.

One profile in general will cover several kilometres, and even larger regions may be covered by such a survey in a relatively short time, giving a map of regional trends, of the depth of the basin and even of smaller structures like salt domes. Among other details these maps and diagrams contain information about the speed of the sound within the various layers of the underground, which are very useful for the interpretation or reflexion seismic surveys.

4.1.6 Small Scale Exploration by Geoelectrics

Geoelectrical surveys belong to the earliest methods of petroleum exploration, but have almost entirely lost their importance because they are restricted to small depths for direct interpretation and because their results mostly become somehow enigmatic when used on a bigger scale.

Their real importance for petroleum exploration is the fact that they gave rise to bore-hole logging some fifty years ago. In modern times geoelectrics are used very rarely, and only in special cases. They are mentioned here mainly for historical reasons.

For some time telluric surveys have been used to outline sedimentary basins lying over crystalline or metamorphous basements. This method, however, takes up too much time, and has been replaced by airborne magnetics and similar methods.

4.1.7 Detail Petroleum Exploration by Reflexion Seismics

Reflexion seismics are the most frequently-used, but also the most expensive of all methods used in petroleum exploration. First and primitive attempts using this technique were made around 1930 and 1935, but nowadays even several variations of the principle are known which may be used for special purposes or which have proved to be better or the best for special regions or under special conditions.

The princple is to send an impulse down into the earth and to measure the time until it comes back as an echo, reflected by petrographic boundaries in the underground.

These impulses may be created by blasting charges of dynamite in shot-holes drilled into the ground down to the water table, by weight dropping, by vibration of heavy masses or similar methods. The acoustic echo is taken up along a line of several hundred metres by 12, 24 or 48 geophones or even clusters of microphones. Every microphone is connected to the registration car by a central cable, and its signals may be recorded on photographic paper or magnetic tape, together with a very fine time scale.

From every shot a series of echos will come from the underground, and the time lapses between them give a rough picture of the structures hidden from the eye or from other methods. Recordings on magnetic tape can be corrected, calculated to real depth and plotted automatically on long profiles by special machines which have been invented to save time and work and to avoid human errors during interpretation. There are some difficult regions, however, where a skilful specialist will reach to better results than even such a very expensive machine.

Such seismic profiles show petrographic boundaries in the underground, or rather boundaries of layers with different acoustic properties. A horizontal boundary between a sandstone and a claystone, or a limestone and a marl will appear in such reports as a more or less broad and more or less conspicuous black bar or a series of black dots or triangles. If the profiles show only a group of nearly parallel lines for every shot, this kind of representation is called „conventional wiggle". Every one of these black bars represents the trace of a small lamp, reflected by the mirror of a galvanometer on photosensitive paper. If the trace of light is sent through a special lens beforehand, all curves will appear as black triangles, called „variable area (VAR)". It is nearly impossible to interpret such a seismic profile geologically without regional experience. In every region there are some disconformities and reflecting horizons with characteristic seismic properties, which may be identified by the shape of their reflection. In general two or three horizons within one tectonic block are sufficient to clear single structures if tectonics are not too complicated. Prospective pays may be identified within such profiles by the distance they should have in that region from the next identified reflector.

A geologist who wants to use seismic profiles for contour maps should know first of all that all depths shown in profiles or seismic maps need not be very exact, because the acoustic speeds usually are not known exactly enough.

That will be possible only if a well has been drilled not too far away, and the acoustic speed has been logged by an appropriate logging system.

Together with the error of depth there may be a certain error concerning the dip of faults, of layers, the relative angle of disconformities and such features.

4.1.8   How to Outline a Saltdome in the Underground

Saltdomes within sedimentary basins will appear in magnetic maps as negative anomalies, because rock salt is not magnetic; they will appear in gravimetric maps as negative anomalies because rock salt has a lower densitiy than all other consolidated sediments; they show high resistivities in geoelectrical surveys; and they show very high acoustic velocities when they are crossed by a profile of refraction seismics. In profiles of reflexion seismics they only show up in a negative manner, that means that all reflectors end along the border of the salt dome or even some hundred metres beforhand. Within the salt mass only confuse, short and irregular reflexions will appear on the profiles.

One of the most important tasks may be to outline such a salt dome as exactly as possible in order to locate a drilling in the highest top of a pay, just before the border fault of the salt dome. Several different methods have been tried in the course of time for such purposes, and some of them have proved to be useful in certain cases.

One of the easiest ways is to run a very close gravimetric survey and to locate the border of the salt mass by exact measurements combined with exact calculation. That method will yield exact results only, however, when the border of the dome is strictly vertical. Irregular borders, which are common in nature, will cause erroneous data and may lead to false drilling locations. One of the best methods known so far is to locate the border of a saltdome by a special arrangement of refraction seismics. This method makes use of the different acoustic speeds of rock salt and of average sediments. If a geophone or still better a chain of several geophones in hanging in a well not too far away from the salt dome, and a shot is fired from a shot-hole on the opposite side of the dome, the acoustic wave front will travel first across the rock salt and then across the sediment until it reaches the well with the geophones. The different times it takes to traverse the salt and the sediment may be calculated easily, and the border of the salt mass may be outlined very exactly. Of course the well or the wells used for such purposes should be as near to the saltdome as possible.

4.1.9   Mapping the Underground by structural Drillings

Sometimes layers with prospective pays and possible reservoirs crop out under some fifty or even a few hundred metres of sediment, but cannot be detected by usual geophysical methods; or a seismic survey would be too expensive and perhaps would not even produce the necessary details. In such cases the best and on the whole the least expensive way will be to drill down to the older layers, draw a core as soon as they have been reached, drill another hole some hundred metres away and continue until the structure of the layers below the quarternary cover seems to be clear, and either a deep drilling may be located or the project may be abandoned.

Such movable drilling rigs, mounted on a lorry or on a sledge, are relatively cheap and may drill a hole of a hundred metres in one day or less, move within a few hours and drill another hole. Care should be taken, however, not to stop a drilling until the core has been dated with certainty, and a second or third core will always be cheaper than a useless drilling with a false or uncertain stratigraphic analysis.

There are some cases where oilbearing structures have been found out by drilling two or three metre holes into the ground by hand. Such drilling equipment, where the drill stems are beaten into the ground by a hammer or drilled into it by a very light and handy motor are sometimes deep enough to penetrate the quarternary cover and to clear structures in the underground.

5   The Practice of Deep Drilling

The location of a deep drilling in unknown regions, called a „wildcat", will usually be fixed more or less exactly on a topo- graphical map by a conference of experienced geologists and geo- physicists. Before the rig may be moved to that location, the exact position must be checked and cleared by the geologist, the driller, the leasing man, and perhaps even by some other specialists. Usually the project is marked by a smaller or bigger circle on the map, with some remarks indicating more or less favorable trends within that circle.

There are some laws concerning the security of such drilling, and other useful facts which even a geologist should know: A drilling location needs a solid and dry underground. Swampy and unreliable grounds require fundaments which are not only expensive but are time-consuming too. If the required location is situated in a swamp, within a group of houses or other similar obstacles, it may often be better to drill a directional hole from the side to the target region, if the distance is not to great. Houses, roads and electric lines are common obstacles in civilised countries. In general a drilling rig must be situated so far from a street, a house or an electric cable, that the top of the rig cannot hit them if it falls down or is blown up by some drilling accident, usually 50 metres. On the other hand a rig will make so much noise, that people in a house only fifty metres away will certainly complain. If they fight it out before a court, they will be sure to win the case and may cause severe difficulties. It is always better to locate the rig a bit further away, or to settle that question before the noise has really begun.

Other things that may cause trouble are roads, zones of water supply, and electrical lines.

A geologist who has to fix a drilling location should bear in mind that the rig has to be moved to that location, that it will be moved on heavy and super-heavy lorries, and that the road leading to the location has to withstand such lorries for a fairly long time. All bridges leading to the location should withstand at least 20 tons, and the road should have a good and solid cover. If not, it will have to be repaired or be covered by steel plates or wooden plates. A short road across a swamp or a swampy creek may cause a delay of several weeks if a temporary bridge has to be built. The quickest, easiest and cheapest way to move a rig to a drilling location is to roll it without unscrewing. Special hydraulic pumps and chain tractors have been developed to carry the rig, and heavy tractors and bulldozers are necessary to move it. Of course there must not be any major obstacle within the moving line, such as bridges, electric power lines, creeks or slopes. On average, only one per cent of rig moving can be done that way. Most times a rig has to be dismantled more or less entirely, and it has to be rebuilt on the location. Modern rigs, therefore, consist of smaller units which may be put together more or less easily and which may be rebuilt within one or two days. Smaller rigs for some hundred metres or even up to some thousand metres are often constructed in a such way that all the rig including the winch, the motors, the mud pumps and part of the tubing may be carried on three or four special lorries only, and may be rebuilt in less than one day. The easiest way to move a drilling rig is by sea, where either the rig alone is shifted for some metres on the platform, or where the entire platform lifts its legs, becomes a swimming object and is towed by some tugboats to the next location, where it sinks its legs down again and turns the platform up to its former position. Even then the underground has to be checked thoroughly as to the solidity of the sediment, the speed of currents and the transport of sediment by the water.

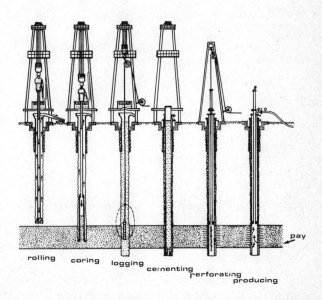

Fig. 42 Some phases of the drilling and the completion of an oil well. The well is drilled down till to the top of the pay formation, using roller bits. The pay formation is cored. After reaching the final depth, the bore hole logs are run and the casing is built in and is cemented. After completion with tubings and wellhead (christmastree), the pay formation is connected with the well by perforations, and the well may begin its production.

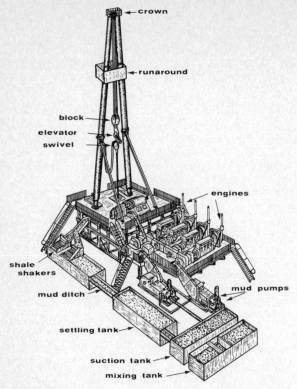

Fig. 43   Schematical picture of a typical rotary rig.

5.1   What a Geologist Should Know about Drilling Technique

Drilling a well for oil or gas means some period of teamwork between the drilling crew, the geologist and his samplers, and some specialists and service people. Teamwork means first of all that every team partner has to understand at least the language of his co-partners, and that he has to know what the others are able to do and what not.

So the geologist has to know the rig, the names of its most important parts, and how they work together. He has to learn something about the difficulties which may occur during drilling, how they may be avoided, and how they may be overcome. Some talking between the driller and the geologist at the right time may save a lot of time and money to the company.

First, however, the young geologist has to learn as much as possible about cuttings and cores, how they are drilled, and how they are brought to the surface. He must be able to reconstruct the layers in the underground by means of the cuttings, to date them, and to run cores at the very moment when they have to be run.

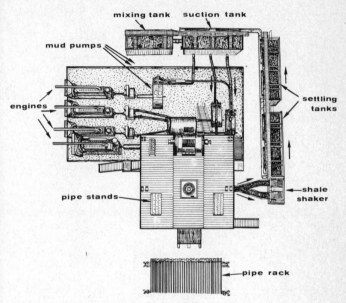

Fig. 44 A similar rig, seen from above.

He has to inspect the cores, to make a rough analysis of their petrophysical properties, and of their pore contents. He has to learn a lot about mud gas logging and borehole logging, on cable tool testing and drill stem testing, and how all these single results of a great variety of methods may be integrated for the purpose of checking the pay and the oil find or field. All that sounds very complicated, and most of these methods come from different parts of natural science, but they may be learned within a relatively short time. They have to be learned indeed if the young geologist, coming from some university, wants to become a good all-round wellsite geologist, and a good team partner to the driller, and later on to the production engineer.

5.1.1   *The Drilling Rig and its Most Important Parts*

Rotary rigs are very similar concerning their principal structure, if
they are to drill for some hundred metres only, or if they are able to
drill some thousand metres. Modern rigs work with jet bits instead of
conventional rotary bits, or even with drilling turbines, but that does
not affect the general construction of the rig, but only the number
and size of engines and mud pumps, and the lowermost part of the
drilling string.

Every typical rotary derrick contains the following main units:

1. Substructure, working platform and mast with crown; the stable
parts which have to carry all other parts;

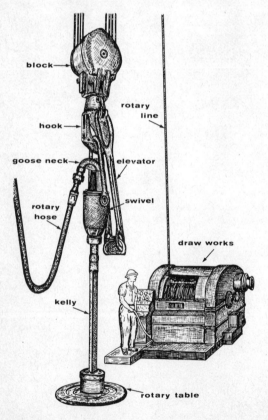

block

rotary
line

hook

goose neck   elevator

rotary
hose   swivel

draw works

kelly

rotary table

Fig. 45   The most important moving parts.

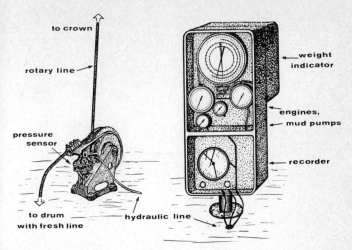

to crown

rotary line

weight indicator

pressure sensor

engines, mud pumps

recorder

to drum with fresh line

hydraulic line

Fig. 46 The drillometer and the sensor for the total weight near the dead end of the rotary line.

2. Rotary table, kelly, drill stems, drill collars, reamers, and rotary bit, jet bit or core barrel; the rotating drilling string;

3. Mud system with mud tanks, stand pipe, rotary hose, swivel, casing, mud funnel, return ditch, shale shakers, settling pit, mixing tank, and slush pumps;

4. Drawworks, rope, travelling block, rotary hook and elevator to move the drilling string up and down;

5. Engines; mostly one engine that drives the rotary table, and several others which drive the slush pumps during drilling and the draw works when the drilling string is run in or out;

6. Blow out preventers with their valves, manifolds and nitrogen containers which are made to shut the casing head in case of emergency within some seconds.

Besides these most important units there are some less important things with special popular names which a newcomer should know:

The *ginpole,* a gallows-like structure at the upper end of the derrick;

the *runaround* with the *ape board,* a balcony-like construction about one third below the top of the derrick, where the lengths of drillpipe are taken out of the elevator and slipped into a rake, or vice versa;

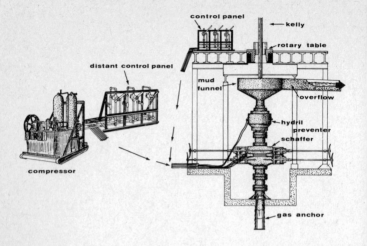

Fig. 47 Blow out preventers and their control panels at the driller's stand and near the rig.

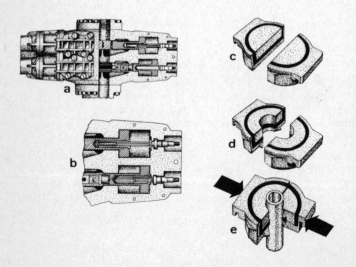

Fig. 48 The Schaffer preventer and its rams which close the bore hole as well around the drill pipe as over the entire diameter.

the *travelling block,* called „Hampelmann" (puppet-on-a-string) by German drillers„

the *swivel* with the *goose neck,* where the rotary hose is fixed;

the *cat heads* on the working platform, which serve to „break", to unscrew the drillpipe;

the *rat hole,* a piece of pipe where the swivel with the kelly is put aside while the drillpipe is run into the hole or vice versa;

the *mouse hole,* another piece of pipe where the next drillpipe is put in, ready for screwing, when the rig is „making hole";

and – most important of all for the well site geologist – the *dog house,* where the driller stands.

Substructure, Platform and Mast

The substructure is not so interesting for a geologist, it is a steel construction which stands on a foundation of concrete and has to carry the derrick, the drillpipe and all other moving and rotating units. Most interesting for a geologist are the casing head in its middle, – the upper end of the bore hole, – the blowout preventers, the mud funnel above them, and the mud return ditch leading to the shale shakers on one side of the substructure.

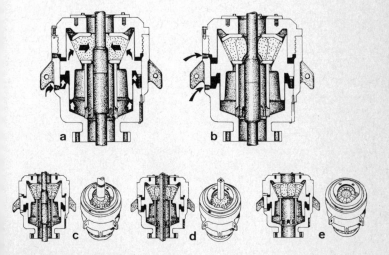

Fig. 49 The Hydril preventer, and how it closes around a tool joint (a), a drill pipe (b), the kelly (d), and even over the whole diameter.

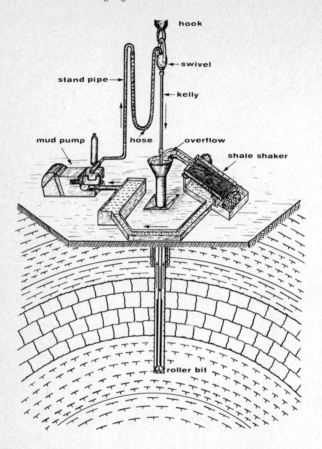

Fig. 50   Schematical flow diagram of the drilling mud system.

The blow out preventers are heavy steel constructions which are able to shut the upper end of the borehole completely within few seconds, and against high pressure. The lower one, called Shaffer preventer, closes the hole by horizontal steel bars which are moved together by compressed nitrogen. One set of these bars may shut the entire cross section of the casing head, the other one has a window which closes around the drillpipe. That means that the kelly has to be drawn out of the preventer before it can be closed. The upper, pot like preventer is called Hydril preventer and is a very complicated construction of steel and rubber elements. It is able to close around the kelly, around the pipe, and to close the entire hole. In addition, most types of Hydril preventers may be rotated with the pipe within the preventer, and the rubber lips closed hermetically around the pipe. These preventers can be shut by a set of valves, which are installed both near the dog house as well as a bit apart from the rig. In case of emergency, when a blowout threatens, the driller can shut the hole immediately. If he is not able to do so in time from the platform, some other roughneck may shut the hole by the other set of valves at a safe distance.

The mud funnel, the mud return ditch and the shale shakers are the most important parts of all the rig for a geologist.

The mud returning from the hole will flow from the funnel through the open channel to the shale shakers, where the solid parts are

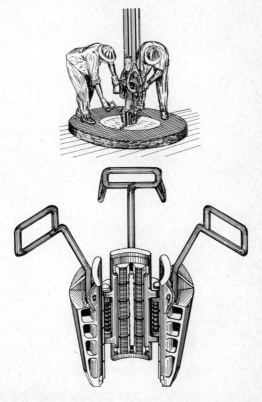

Fig. 51   Roughnecks running the slips into the rotary table in order to block the drilling string, and a set of such slips. Most modern rigs have slips which may be set and withdrawn automatically.

Fig. 52 Roughnecks unscrewing part of the drilling string using rotary tongues and one of the catheads.

sieved off. The open channel between the funnel and the shale shakers is the right place for mud gas logging. Usually a small float is attached to the walls of the channel, which bears a mixer and a gas dome, where the gas is sucked out of the mud and passed to the gas detectors in the gas lab.

The shale shakers are the place where the cuttings coming from the borehole are sieved off. A part of them is taken away for geologic analysis. Most cuttings are carried away to the mud pit, where they are dried and buried, together with part of the drilling mud.

The Platform

The drilling platform is the real working site to the roughnecks. The round steel construction in its middle is the rotary table, one of the most important inventions of drilling technique. Usually it is driven by a special engine, and it has to rotate the entire drilling string. This is done by the kelly, a four- or six-sided steel bar, which hangs on the swivel. The sides of the kelly fit into recesses of drive bushings of the rotary table. If the table rotates, it turns the shaft of the kelly and the kelly is screwed to the upper end of the drillpipe which is turned likewise. The upper end of the kelly is attached to the swivel, a high pressure stuffing box. The mud, coming from the slush pumps, the stand pipe and the rotary hose, enters the swivel by the goose neck, and from there it enters the hollow shaft of the kelly, the drilling string, and goes to the bottom of the hole.

The swivel is suspended by the travelling block, the rope and the crown on top of the derrick. It is kept suspended by the rotary hook, and so it can be moved up and down by the drawworks. When a new pipe has to be added to the drilling string, or when the string has to be run out of hole, the kelly is screwed off the drillpipes and put aside into the rat hole. The drillpipes are kept then in the rotary table by slips, slip rings or power slips. The elevator fits around the joints of the drillpipe, it is taken to move the drill string up and down. When „hole is made", i.e. when the well is drilled deeper, the kelly can slip downwards in its buschings until the swivel is just above the

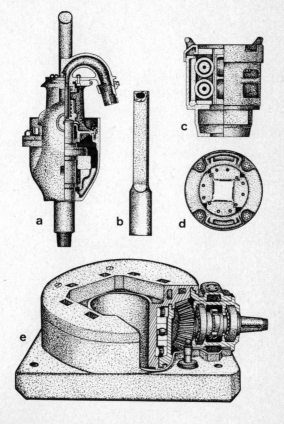

Fig. 53 Cutaway view of a rotary table (e), of a swivel with the goose neck (a), a kelly (b) and of the bushings (c, d).

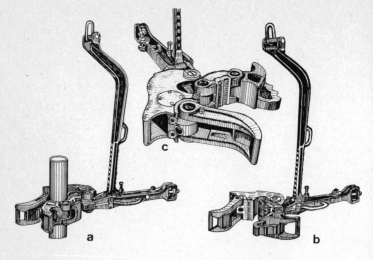

Fig. 54   Rotary tongues open and gripping around a drill pipe.

rotary table. Then rotation is stopped, the entire string is drawn up until the uppermost drillpipe appears above the rotary table, and the slips are run into the recesses of the rotary table instead of the bushings. Then the kelly is screwed off and put aside into the rat hole.

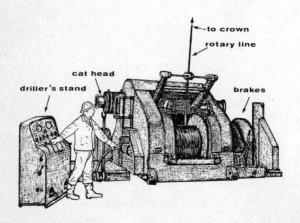

Fig. 55   The driller's stand with the control panels, the drawworks, the brakes, the cat heads, and the rotary line.

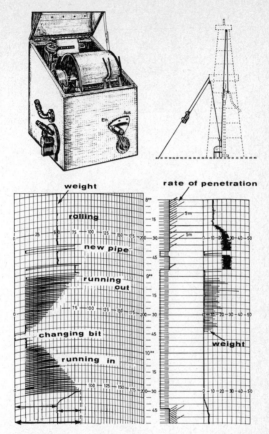

Fig. 56 A drill recorder and its diagram, showing the total weight, the weight on the bit, and the rate of penetration.

A new drillpipe has been put into the mouse hole before; it is hung into the elevator, the winch draws it out of the hole, and then it is screwed into the top of the drilling string. Then the slips are drawn, the entire string is lowered into the hole, the slips are run again, the kelly is hung in the rotary hook, it is drawn out of the rat hole, screwed on top of the uppermost pipe, and then drilling may continue until the swivel arrives again over the rotary table. Such drillpipes have a length of about ten metres, and three of them together form one „length". If the whole string has to be drawn up, because the bit is worn, or because a core has to be drilled, the procedure is just the reverse of „running in". The kelly is put aside, and length

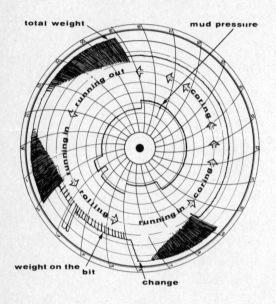

Fig. 57 Dial and diagram of a regular drillometer, showing the weight on the bit, the total weight, and the mud pressure.

after length are screwed off. The lower end of every length is moved aside to its place on the platform, and the upper end is hung out of the elevator by the man on the ape board, and shifted into its place in the rake.

The pipes at the bottom end of the drill string are thicker and heavier than the normal drillpipes, they are called drill collars, and have to keep the hole straight, together with the reamers.

The bit on the lower end of the drill collars is the most important drilling tool. It consists of a shaft and three or even four rollers with longer or shorter teeth. The teeth are made from extremely hard steel, or even of tungsten carbide. In general there are three types of teeth, long ones for soft formations, and short ones for hard formation. For extremely hard formation and for turbine drilling special bits are used without rollers, and studded with diamonds.

The teeth break the cuttings loose from the bottom of the hole when the bit rolls around, and the mud stream takes them up and carries them to the surface.

The mud outlets between the rollers are most important for drilling, they are fitted with jet nozzles, and the jet stream helps to „make hole".

Another important feature on the drilling platform is the driller's stand, the „doghouse". From there all engines, the drawworks with its brakes, the slush pumps and the blow out preventers can be operated, and their function can be controlled by the drillometer and other control units.

When the rig is „making hole", the driller has to actuate only the brakes which keep the drilling string suspended. The bit would not turn if all the load of the drill collars and the drillpipe were resting on it, part of that load has to be kept suspended by the drawworks via travelling block, swivel, rope, and crown block. The „differential pressure" which rests on the bit is still high enough; it amounts to

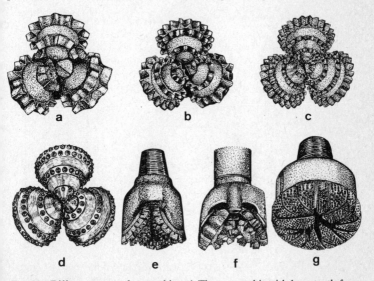

Fig. 58 Different types of rotary bits. a) Three-cone bit with long teeth for soft formation. b) Three-cone bit with medium teeth for solid formation. c) Three-cone bit with short teeth for hard formation. d) Carbide insert or "button" bit for abrasive formation. e) Three-cone bit with conventional mud-discharge ports. f) Three-cone bit with jet nozzles. g) Diamond bit.

fifteen tons and more, depending on the type of rig and the type of formation. The driller has in front of him the drillometer, which shows the „total load", which is hanging on the travelling block, and the „differential load", which is forcing the teeth of the bit into the formation.

From time to time the driller will lift the string until the first connection appears above the rotary table. He wants to keep the bottom of the hole and the drill collars clear of cuttings, and regular lifting ensures that the string doesn't stick in the hole. Alle these movements are recorded on the chart of the drillometer together with the mud pressure and other details. The drillometer itself has its sensor on the lower end of the rope, the so called dead end. A pressure cell measures the tension of the rope, and the signal is led to the drillometer by a hydraulic line. The pentration rate usually is logged by means of a steel rope inserted at the swivel which drives the pen of the recorder, while the chart is moved forward by a clock-work. The chart showing the penetration rate is most useful for comparing the hardness of the formation, but most recorders have their chart arranged in a way that it is a very disagreeable and time-consuming work to produce a curve which shows the depth of the hole versus penetration rate, the only interesting relationship for a geologist. This curve would be a very useful base for gas logging, too, and it could be transmitted electrically to the mud gas recorders.

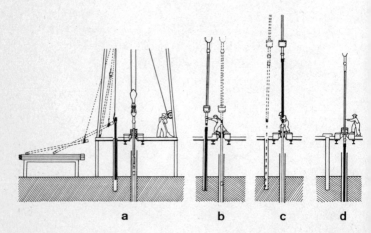

a   b   c   d

Fig. 59 How a new length of drill pipe is added to the drilling string. a) The new length of drill pipe is run from the pipe rack into the mouse hole. b) The drilling string is hauled up and blocked by the slips in the rotary table. The kelly is screwed off and is connected with the new pipe. c) The kelly with the new pipe is hauled up and gets connected with the drilling string. d) The slips are withdrawn, and drilling may rebegin (after Le Roy).

When the drilling string is run in or out, the driller and his men have to work very hard. The driller himself actuates the drawworks and the brakes, he moves the travelling block up and down, with pipes hanging in the elevator. One man is working on the ape board, taking the lengths of pipe out of the elevator and putting them into the rake or vice versa, and two men are busy to run the slips into the rotary table, to screw and unscrew the lengths of pipe and to move them to or from their stand. If the hole is several thousand metres deep, such a „round turn" with running in and out, – usually in order to change the bit – may last many hours or even an entire day, and it is hard and dangerous work. A newcomer should keep off the platform then, or at least remain in a safe corner, where he is out of the reach of splashing mud, he should keep away from the catheads with which the rotary tongues are actuated, and far away from the rotary table.

The Mast or Derrick

The huge mast of a drilling rig is its most eye-catching part, but not so interesting for a geologist. The mast has to carry all the load that hangs on the travelling block, and the drilling capacity of a rig is limited by the maximum crown load of the mast, usually a very impressive figure. Old fashioned derricks consist of a very solid construction of innumerable triangles, which are screwed together, and have to be unscrewed and screwed again when the rig has to move from one well to the next. Other types of derricks consist of four piles, which carry the runaround and the crown. The piles consist

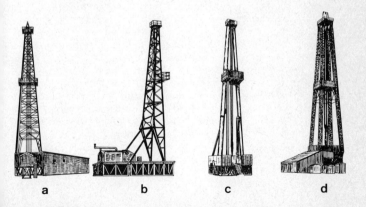

Fig. 60 Different types of derricks. a) Old lattice-type or pennsylvanian derrick. b) Self-erecting rig of the lattice-type. d) Self-erecting rig consisting of two steel tubes. d) Heavy duty derrick consisting of four piles.

Fig. 61   So called self-erecting types of rigs during erection. The masts are screwed together on the ground, and they are drawn up using the rotary line, the drawworks, and the engines.

of four or more parts each, which can be put together more easily, and in a less time-consuming way. Other types of masts show only two piles or tubes. They may be screwed together while they are still lying on the ground, and they are drawn into upright position by the drawworks and the rope of their rig. It would take up too much space to mention all variations, and the technical advantages and disadvantages of all types of derricks which have been constructed during the long history of drilling technique. Interesting for a geologist are some types of small rigs, mounted on a trailer, which are used for structural drillings, for drilling shot holes for seismic exploration, or similar purposes. These types usually have a telescopic mast which can be turned upward for drilling, and the whole unit may change its place to drill another hole within a few hours or even less. There are some bigger types of such trailer-mounted rigs which are even able to drill more than one thousand metres of slim hole, but usually such types experience difficulties if casing has to be run into a deeper hole.

The Mud Circulation System

Fresh mud is mixed in special tanks according to the recipe given by the specialist, and under the surveillance of an experienced man. For a deep drilling of about 3,000 metres several tons of chemicals will be used, and every day several cubic metres of water are necessary for mixing. Mud types containing montmorillonite or attapulgite need some time for soaking and can be used only after one day or longer of preparation.

From the mixing tanks the fresh mud can be pumped into circulation. The mud pumps move it to a vertival pipe half up the derrick, where it enters a high pressure rotary hose leading to the goose neck on the swivel. From the swivel the mud enters the drilling string by the kelly and goes downhole to the bit where it is pressed through the

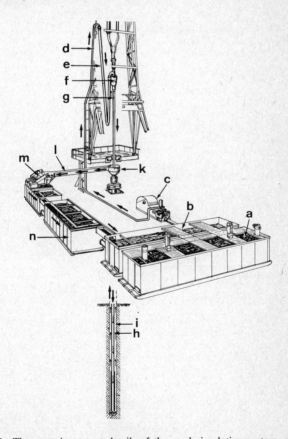

Fig. 62 The most important details of the mud circulation system. a) The mixing tank, b) the suction tank, c) the mud pumps, d) the mud pipe, e) the mud hose, f) the swivel with the goose neck, h) the drilling string, i) the bore hole with the first casing, k) the mud funnel, l) the overflow, m) the shale shakers and n) the settling tanks with the mud return ditch.

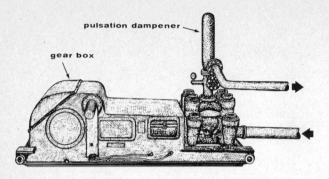

Fig. 63   A duplex mud pump.

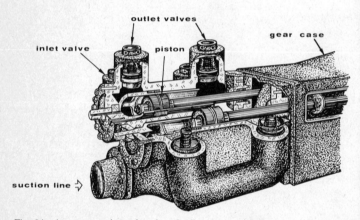

Fig. 64   A cutaway view of such a duplex pump with its valves, pistons and cylinders.

jet nozzles and erodes the bottom of the bore hole. Together with the brickles of rock, which have been broken loose by the bit and by the mud jet, the mud then returns to the surface by the outer rim space of the hole, passes through the blowout preventers, enters the mud funnel, and leaves it by the overflow.

Within the overflow one of the most important geological control instruments is floating on the surface of the mud, the stirrer and sucker of the mud gas logging system. Any gases or oil traces contained within the mud are sucked here into a pipeline leading to mud gas detectors within the geological laboratory.

Just after the overflow channel the mud stream falls on the screen of the shale shakers, where all solid particles bigger than two or three millimetres are sieved off, and where the samples for geological analysis will be taken.

The mud itself falls through the sieves and returns to the suction tanks by a long return channel, where solid and heavy parts may still settle down and can be removed from time to time.

Any sand contained within the mud will destroy the slush pumps, the mud lines, the drill string and may cause a lot of damage by its high speed. In formations containing much loose sand, the sand has to be removed from the mud by so-called desanders.

These desanders work like a cyclone used for separating minerals. The entire micture is brought into rotational movement, and the heavier particles like sand grains are thrown to the outer sides of the cyclone funnel, where they pass down to the outlet and may be taken off into a mud waste tank. The lighter part leaves out by the central pipe and may be returned to the mud stream.

Sometimes the mud may take up so much gas from gas bearing formation, that the mud becomes too light and could be blown out by the formation pressure. Weighting the mud by baryte and other heavy minerals will not always be very successful, because weighted mud needs higher viscosity to keep the baryte floating, and with higher viscosity a good part of the gas bubbles will remain within

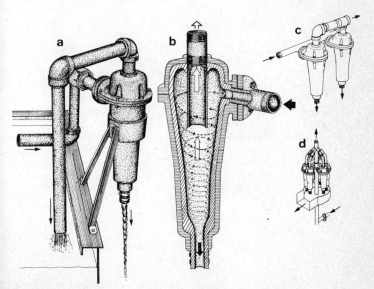

Fig. 65 The principle of desanding. A single desander and groups of them.

the mud, will be pumped a second time into the hole and will increase the danger of blow outs.

The gas bubbles may be sucked out of the mud by a so called degasser, where the mud is spread out over screens and overflow plates, while a vacuum pump creates a certain underpressure. Using a degasser in time may save the mud from weighting, and baryte will not only cost money, but will certainly damage the pay, and that kind of damage cannot be repaired by acidizing.

From time to time, part of the mud has to be taken out of circulation because if has taken up too much debris from the formation and will no longer fulfil the routine tests for water loss, viscosity, density etc.

So nearly every day one or more cubic metres are pumped into the mud pit, together with the coarse material from the shale shaker. These mud pits are a constant source of trouble concerning pollution of ground water with mud chemicals, especially oil and salt. They have to be watched, therefore, to avoid overflow, and they have to be planned carefully to avoid long-lasting damage to trees, plants and even animals in the surroundings. One should consider that cows, sheep and many kinds of deer are extremely fond of salt and could try to lap it up from salty mud. They can be poisoned then by chemicals contained in the mud pit.

5.1.2   What a Geologist Should Know about Drilling Fluids

Nearly everywhere on and around the rig there are pipes, ditches, tubes and tanks containing a grey, brown or even red kind of pudding, called drilling-fluid or simply mud. This mud system is extremely important for „making hole", and it is a most important source of information for the well site geologist and his samplers. It brings the cuttings of the formation to the surface, and it will contain oil droplets or gas bubbles if the hole is drilled into oil- or gas-bearing formations. If the hole is drilled into a formation containing rock salt, the salt will go into solution, and the electrical resistivity of the mud will drop rapidly. If potassium salts are drilled through, they may be detected in the mud by nuclear counters, because part of the potassium salt is radio-active ^{40}K, emitting gamma and beta rays.

Such drilling muds are very complicated chemical and rheologic systems, ad it would fill more than one special text book to mention all recipes for all types of mud, and their technical, chemical and rheologic properties. A well-site geologist should know at least some important features about the type of mud which is used to drill the well, because not only the mud gas logs, but even the possibilities of oil and gas detection within cuttings and cores may be affected.

On the other hand there are many oil companies where the properties of the drilling mud have to be controlled by the samplers in the geological lab, and where even routine mixing recipes are given by that personnel. In general, however, mud mixing has become a real and complicated sciene, and high penetration rates and stable holes with a good calibre are the result of good mud practice.

The most important characteristics of drilling-fluids are:

1. viscosity, gel strength, and colloidity,

2. density or contents in solid material,

3. fluid loss,

4. sand and silt contents,

5. salt contents, especially contents in potash salts.

Viscosity, gel strength and colloidity are the result of the contents of montmorillonite, attapulgite, CMC, gypsum, spersene, quebracho, and similar material. Every type of drilling-fluid has its special recipe to create the pudding-like structure which is necessary to carry the cuttings out of the hole, and to avoid the contamination of the bore hole walls with water from the drilling mud. It would take up too much space to copy here the recipes of only the most common types of mud; newcomers interested in such material should use special text books, or consult the central mud lab of their company.

Routine control of viscosity is made by the use of a Marsh funnel, or better by a Stormer or Fann viscosimeter.

The Marsh funnel is filled up to its upper rim with mud, while one of the operator's fingers keeps the outlet closed. Then the finger is taken away, while the button of a stop watch is pressed. The time which is necessary for the mud to run out of the funnel is in some way related to the viscosity and the colloidity of the mud.

The viscosimeters have to be kept at a constant temperature, if the results are to be valid. They usually contain a rotating cylinder which is connected to its axis by a spring system. The torque between the axis and the cylinder bears a close relation to the structural viscosity of the mud. The viscosity of most drilling muds may be regulated by special additives, such as a special hydrophosphoric salt of soda. If the viscosity is too low, some more montmorillontie, attapulgite, CMC, gypsum, asphalt powder, or similar material has to be added.

The density of the drilling fluid is very important to keep the formation fluids under control, and to avoid blowouts as well as severe fluid loss. Water, oil or gas in porous or vugular rock usually are under a certain static pressure, which corresponds to the weight of a column of salt water, reaching from that formation up to the water table just below the surface. This so called hydrostatic pressure equals 1.1 atmospheres per ten metres, and the drilling fluid must have a higher density to keep the pore contents under control. Most drilling muds, therefore, have a density of 1.3 or 1.4. In some regions, however, especially in young geological basins with poorly consolidated sediments, or near salt domes which are in a state of movement, such pore fluids may show severe overpressure, and bore holes in such regions have to use weighted drilling muds. The most common materials used for weighting are calcite powder, baryte powder, or even hematite or galena. Baryte, hematite and galena, however, are insoluble in common acids, and they may block any pay when the well is to be taken into production. Calcite, on the other hand, may be removed from the bore hole walls easily by any type of acid frac, and the type of weithing material should be considered thoroughly before it is used. Weighted muds, in general, have to be run at a higher viscosity than normal ones, and usually keep every trace of gas inclosed for relatively long time, so that the real weight of the mud has to be controlled by a degasser, a big machine which sucks the gas out of the mud by underpressure.

The density control of drilling fluid is effected by a special mud balance, or better by an areometer. The solid contents may be controlled by the same method which is used to control the fluid loss. A certain quantity of mud is pressed through a filter, and the thickness of the filter cake can then be measured with a caliper square, or checked on a balance after washing and drying.

The *fluid loss* of a mud mixture usually depends on its density. If the mud column is heavier than necessary, some mud will be filtrated into pervious formations, and the solid parts will cover the bore hole walls as a mud cake. These filtrate losses, however, are usually small, and of minor importance. In vugular formations, or in hard rock with abundant fissures, however, the mud may run very quickly into such flowing channels, and get lost. If the slush pumps will not fill the lost fluid up quickly enough, or if not enough mud is ready for use, the mud level may drop to a depth where the counterweight against the formation pressure is no longer present, and the bore hole contents may blow out.

Every driller, therefore, has to look for a reliable mud level recorder, that gives alarm as soon as the mud level disappears below the casing head. Such recorders even give alarm when the mud level drops because the drill pipes are run out of the hole, but even that may cause a blow out if the hole is not refilled in time.

If the mud level really drops, and bigger quantities of mud disappear in fissures or somewhere else, there are some special additives which are believed to stop the fluid loss quicker than ordinary mud. In practice the best additive for fluid loss is that which is available without any loss of time, like sawdust, micas, dried potatoes, konfetti made from plastik material, or similar stuff. In most cases, however, it may be sufficient to reduce the density of the drilling mud to the lowest level possible, just down to the density which is absolutely necessary to counterweight the formation pressure.

In routine work, the fluid loss of drilling mud is ascertained by the brass ring method and by the filter press method.

For the brass ring method, a filter paper of constant quality is put on a glass plate, the brass ring is laid on the middle of the filter paper, and then the ring is filled with mud up to its upper rim. Some of the water will be sucked out of the mud after some time, and the speed with which the border of the water creeps has to be measured by a stop watch. The filterpress method, or baroid press method, is effected by a special filter press. The mud container of the press is supplied with a filter paper, then it is filled up with mud, it is inserted in the pressure system, and the mud is run through the filter with compressed air from a steel bottle. The quantity of water passing through the filter is recorded against time, and after the test, the mud cake on the filter may be measured to ascertain the solid contents of the mud.

The mud filtrate, of course, should be used to ascertain the salt contents, using a conductivity cell and a measuring bridge.

The *sand and silt contents* are not only interesting from the geological point of view, but still more for technical reasons. Any sand grains within the mud will cause severe damage to the mud pipes, the drill pipes, and the slush pumps in the course of time. Any greater contents of sand or silt have to be avoided therefore, and in regions with numerous sandy formations, the sand contents have to be checked.

The most efficient check is to wash a certain quantity of mud through one coarse sieve, which keeps the real cuttings back, and a second fine sieve which lets only the real mud contents pass. The sand, remaining on the lower sieve, is rinsed, dried and weighed on a balance, or measured in a gridded glass cylinder.

If the sand contents are too high, part of the mud can be put aside and be pumped to the mud pit. In regions with several sandy formations this method would be too expensive. The mud has to be treated with so-called desanders, which work after the principle of a cyclone. The mud enters a desander from the outside, gets into rotation, and the heavier sand grains are thrown to the outer edge of the funnel. The sand may be drawn off at the lower end, while the mud enters the central outlet. Sometimes entire batteries of such desanders have to be used to keep the sand level down, and to reduce the damage to drillpipes and slush pumps.

The *salt contents* of drilling muds are important for the stability of the mud, and likewise for the stability of the hole. On the other hand, any salt contents are interesting for the geological analysis of the bore hole, because they may be useful to identify special salt-bearing formations. Some bore hole logging systems are sensitive to the conductivity of the mud, and of the mud filtrate in porous formations. Most types of mud are sensitive to salt because their colloidal structure may be destroyed by ion exchange, especially montmorillonite and attapulgite muds.

Usual laboratory equipment for the control of resistivity will not work well enough in drilling muds because of their high contents of solids. Any measuring cells or probes therefore have to use the so-called four-point-method, using two outer electrodes which supply the electricity, and two inner electrodes for measuring. In principle any conductivity bridge may be used, which fits four electrodes. Streaks of potash salts in salt-bearing formations may cause severe cavings and may endanger the entire bore hole. The only way to detect potash salts in practice is to ascertain the nuclear gamma or beta activity of the drilling-mud, because potash salts contain active ^{40}K, which may be detected by a scintillometer counter or a proportional counter.

The gamma counter is not affected by the steel channel of the mud outlet and can be fixed on its side. Beta counters are very sensitive to every kind of shielding, and any proportional counter used for such a purpose should be kept floating on the surface of the mud.

5.1.3 Blow Out and Lost Circulation, a Mud Engineering Problem with some Geological Points of View

Blow outs may cause the loss of the entire rig, the loss of considerable amounts of oil and gas, and even the loss of lives. In general they are disastrous in every respect, and care should be taken to avoid them if possible. Blow outs during drilling may occur when the mud has become too light to withstand formation pressure. In general the formation pressure may be calculated easily, and the weight of the mud may be kept high enough to withstand it, but there is no rule without exceptions.

In regions with young, still unconsolidated sediments there may occur formation pressures which are much higher than foreseen. The geological reason for such high pressures are quite simple to understand: any layer of young clay will give off considerable amount of water until it becomes a solid claystone. This water will be filtrated into the next layer of sand, and these layers of sand sometimes have a restricted volume without direct outlet to another sand. On the other side these sands are not yet consolidated and will continue to loose some pore space until they become solid sandstones, which means that they may contain a higher pressure in their pore space than they

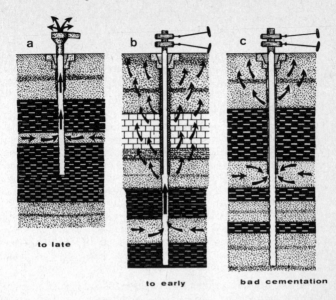

Fig. 66 Blow outs and their geologic reasons. a) The well hit a pay with gas or oil before the gas anchor had been run in. b) The gas anchor has been run in though there are no tight formations which could seal it. c) The gas anchor has been run in correctly, but its cementation is bad. Gas penetrates into the annulus between casing and formation and blows out through the porous surface sediments.

should have. On the other hand, regions which are sinking down in geological times often preserve the former, lower pressure of their higher position in their pore space and so show a lower pressure than they should have according to their depth below the water table.

Such facts usually show up only with the first wildcat drilled down in such a region, and most blow outs occur with such wildcats. Of course there are blow out preventers, but sometimes such eruptions occur so suddenly, that the preventer cannot be closed quickly enough.

Even if the preventer has been closed, the gas anchor may give way, or its cementation may be imperfect. Gas and water creeping up between the wall of the hole and the casing wall may wash away the fundaments of the rig, gas may begin to burn by some accident, and usually it will take a lot of time to stop such a wild flow. So the first geological task is to put the gas anchor deep enough, to have enough metres of solid and tight rock for a good cementation, but not so deep that a gas eruption could occur before the gas anchor is run and has been cemented. A geologist should know that water with some dissolved gas may become much more dangerous than a regular gas field, and that the mud level has to be watched thoroughly when sandy formations are drilled. Gas bubbles in the mud stream, and mud flowing over the borders of the mud return channel show that danger is not far off, and that the mud has to be conditioned somehow. Care should be taken, however, to check the type of gas by the mud gas detector, because a slush pump with a rotten seal may pump the mud full of air and cause a similar effect.

On the other hand it is not always possible to drill with heavy mud only, because that would cause water loss, and heavy water loss may cause blow outs if the mud column within the hole sinks down without being refilled quickly enough. Every porous formation will cause a certain water loss by filtration, but this water loss is stopped after a short time by the mud cake. Most water loss is caused by vuggy and fractured formations, especially limestones and dolomites. Cracks and fissures in hard formations may take up enormous quantities of fluid within a very short time, and if any gas or oil pay is open to the bore hole at the same time, the well may blow out without any warning, if there is no mud level recorder, which gives an alarm when the mud level begins to drop.

Water loss will not be so high when light muds are used, and it may be countered by adding sawdust, dried potatoes or similar material to the mud, until the fissures are blocked, and mud returns to the surface again.

The ideal type of mud, therefore, should be heavy enough to keep the reservoir fluids within the reservoir, but it should not be so heavy that is causes water loss.

It should have a good viscosity to carry the cuttings out of the hole, and to keep them suspended when the circulation is stopped. On the other hand the viscosity should not be too high in order to avoid gas cushions. These conflicting requirements show why it is so difficult, – or rather impossible – to develop one ideal type of mud, and why constant checking and constant mud engineering are necessary if a good penetration rate without trouble is to be achieved.

The best evidence for good mud engineering is a good calibre diagram when the bore hole logs are run, with small mud cake covering the porous parts and no or almost no cavings in shaly formations. Good mud practice will avoid skin effects when the pay is taken into production, and it will help to get a good base for cementation and completion.

6 Mud Gas Logging

Mud gas logging has become a real science during the last years, and a complete representation of its chemical and physical fundamental principles and its different methods would fill an entire volume. In the beginning the only aim of mud gas logging was to warn the driller against possible blow outs. Meantime mud gas logging has became a valuable geological method of pay detection during drilling, and it has been combined with a thorough analysis of cuttings. All that is done in a small laboratory on the well site, provided with water, electricity, with the necessary detectors, and geological and chemical equipment.

The most important part of the mud gas analyser is the so-called „float" in the overflow channel near the mud funnel. This float contains a stirrer, which is driven by an electric motor, and a suction line which draws the gas off and leads it to the laboratory. This float must be kept clean, free of clogging, and in good contact with the mud returning from the bore hole. If the float runs well the detection of gas is only a question of routine.

The gas coming from the float may contain methane, higher hydro-carbon gases, atmospheric air, carbon dioxide, hydrogen sulphide or even only nitrogen. It seems clear that several different detectors must be used if all these different gases are to be logged. In practice it depends on regional experience what detectors are used best, and what types of gas should be recorded on the chart.

6.1 Cold Wire and Hot Wire Detectors

These two types, sold by numerous companies, are similar in shape and name but very different in method.

Cold wire detectors contain two filaments of platinum, shielded by glass or a plastic vernice. One of these filaments is placed in the gas stream to be detected, the other in atmospheric air. The two filaments represent two branches of a Wheatstone bridge, and any difference in gas composition between the gas stream and the atmospheric air will cause a difference of conductivity of the platinum filaments, and a corresponding cross potential. Such detectors are very good if only methane is to be expected, but no hydrogen or carbonic acid.

Hot wire detectors contain bare platinum filaments instead of shielded ones, and these filaments are heated electrically to about 300 centigrade. Any hydrocarbon gases will burn on the gas fila-ment, increasing its temperature and creating a cross potential. Such detectors are very useful for simple cases too, but they do not

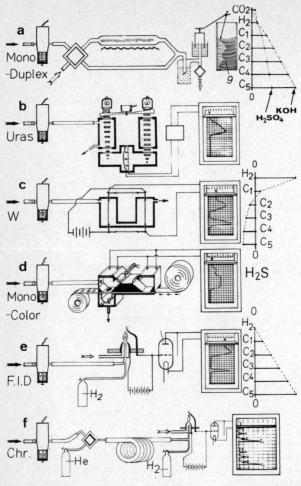

Fig. 67 Some different types of gas detectors. a) Mono-duplex detector, where hydrocarbon gases are burnt in an oven, and where the difference of volume is recorded. b) Uras gas detector, a simple type of infrared spectroscopy. One type of gas or two types of gas may be recorded specifically. c) A hot wire detector, utilizing the different thermal conductivity of gases. d) Monocolor detector for the detection of sulphuric gases. A stripe of paper containing a lead salt becomes dark or even black when it gets into contact with sulphuric gases. e) Flame ionisation detector for detecting hydrocarbon gases. Such gases become electrically conductive at the high temperature of the flame. f) Gas chromatograph, using molecular sieves to separate the different components of a gas sample.

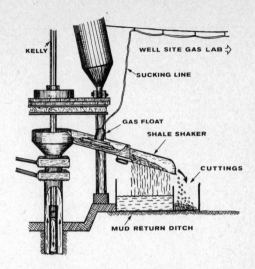

KELLY

WELL SITE GAS LAB ⇒

SUCKING LINE

GAS FLOAT

SHALE SHAKER

CUTTINGS

MUD RETURN DITCH

Fig. 68 The overflow of the mud near the well head with the gas stirrer and the mud gas suction line.

178378

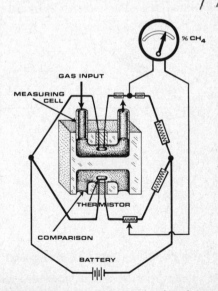

% CH$_4$

GAS INPUT

MEASURING CELL

THERMISTOR

COMPARISON

BATTERY

Fig. 69 Schematic circuit of a gas detector using thermistors. One thermistor is kept within atmospheric air, the other one gets into contact with the gas sample. The working principle is similar to that of the hot-wire detector.

allow any geological conclusions to be drawn unless only one type of
hydrocarbon gas occurs, and probably they will be replaced by
pellistor sensors in future.

6.2   Flame Ionisation Detectors (F.I.D.)

These detectors have a little flame burning which is fed from a steel
bottle containing pure hydrogen. The mud gas is fed into the flame,
and a current of some 50 or 100 Volts DC is fed to the burner and to
a small screen made of platinum within the flame. Usually no current
will flow between the burner and the screen. If the gas stream contains hydrocarbon gases, however, or other gases containing carbon,
the flame will become conductive, and the total contents of carbon
gases may be logged. These F.I.D. are very rugged and very reliable
instruments, they are not sensitive to water vapour and therefore
they have replaced the wire detectors which show a certain sensitivity
to vapour.

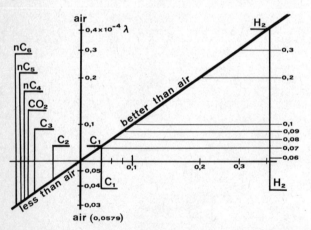

Fig. 70 The thermal conductivity of some natural gases. The diagram shows
that hydrogen has much more influence than methan, while the heavier
hydrocarbon gases have just an opposite effect.

6.3   Surface Active Transistors Used for Gas Logging

For some time some types of transistors have been used for direct
gas logging which show different electric characteristics if their
surface is in contact with gases or not. In general they seem to be
sentitive to a very broad spectrum of gases, and so they may be used

for very general purpose only. Up to now only very little is known about their qualities, and only the future can tell if they are really useful for gas logging, and for what types of gas they may be used best.

6.4 Pellistor Detectors

For some time detectors of a new type have been on the market, which make use of so-called pellistors. They are small pellets of ceramic material, containing a filament of platinum. One of two pellistors always contains a surface layer of some contact material. Both pellistors are put within the gas stream, and form branches of a Wheatstone bridge like the filaments of a hot wire detector. If the gas stream contains any gases which are subject to oxidation, they will be burnt within the surface contact material of one of the pellistors and heat it up, while the pellistor without contact material will remain cool. Of course such a detector will log carbon monoxide as well as hydrocarbon gases, hydrogen sulphide and others. Care should be taken therefore, to keep the mud logging system and even the mud system free of engine exhaust gases, lubricating oil and similar substances.

6.5 Sulphide Detectors

If hydrogen sulphide is to be expected, or shows up with its characteristic ball smell, it can be detected by a simple chemical reaction with salts of heavy metals like lead or iron. If the gas stream is blown against a stripe of paper impregnated with lead acetate, for example, it will be blackened by the development of lead sulphide. This blackened stripe is compared electrooptically with the white stripe, and the difference in brightness will show up as a corresponding cross potential of a measuring bridge. Other sulphide detectors make use of physicochemical, reversible reactions creating potential differences within a special measuring cell.

6.6 Infra-Red Detectors

Infra-red gas detectors make use of the well-known fact that every gas will absorb special bands out of infra-red light. In practice there are two tubes within the detector, one filled with nitrogen or another inert gas, and one with the mud gas stream. The gas which is to be logged is contained in another tube below the two test tubes. This lower measuring tube is separated by an electric pressure sensor into two compartments. If infra-red light is led through the two vertical tubes, the nitrogen will filter out the nitrogen bands of the

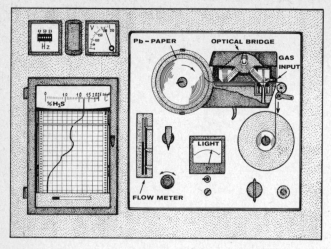

Fig. 71 Cutaway view of a detector for hydrogen sulphide. The gas is blown against a stripe of paper containing a lead salt. The stripe turns black if its gets into contact with hydrogen sulphide, and an optical bridge compares the white stripe before and after the test.

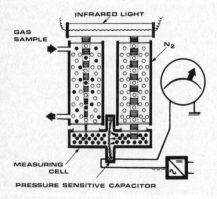

Fig. 72 Uras gas detector using infrared light for the specific detection of one single type of gas.

light, and the gas stream will filter out bands of gases contained in it. If the gas to be logged is methane for example, the measuring tube will be filled with it. The light bands of methane will then pass unfiltered through the nitrogen, but if the gas stream contains methane, that part of the measuring tube will become colder, and the pressure within that compartment will drop. These differences in pressure will be logged and recorded on a chart. Of course every type of gas may be logged, and if another system with a third measuring compartment is added, two different gases may be logged at the same time.

This may become important if the geologist has to predict if the pay contains pure gas, or condensate, or oil. Gas fields usually show only methane in their mud gas, condensate fields show a higher amount of propane, and oil fields a very high pentane/methane ratio. Such ratios, however, depend very much on the type of drilling mud

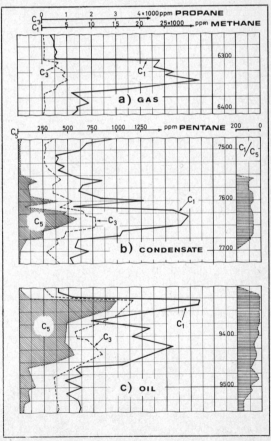

Fig. 73 Diagrams of a mud gas survey in wells with natural gas (a), condensate (b) and oil (c). Higher parts of heavier hydrocarbon gases point to oilfields or condensate fields. The most striking differences are shown by the ratio of methane versus pentane.

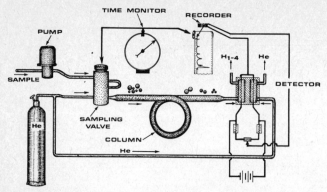

Fig. 74 Schematical circuit of a gas chromatograph. The sample is pumped into the detector by a sampling valve. Thé column works like a capillary sieve and separates the different gases which are washed out of the column by helium gas. The final detector is of the hot-wire or of the flame ionisation type.

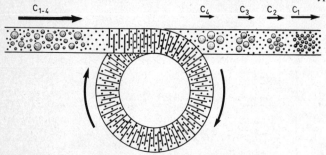

Fig. 75 Schematical mode of operation of such a column. The sample containing several different natural gases is separated into its components, which arrive at the final detector successively, the smaller molecules at first.

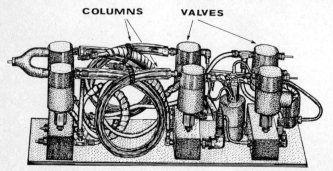

Fig. 76 The real appearance of such a gas chromatograph.

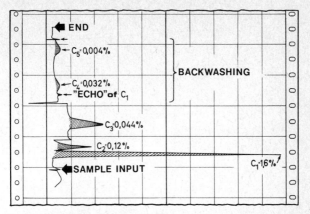

Fig. 77   Diagram of a mud gas sample from a gas chromatograph. Every type of hydrocarbon gases produces a more or less distinct peak. After the end of the analysis, the direction of flow within the column is reversed, and the last traces of gas are washed out of the molecular sieve.

and on its weight and its viscosity. It should be considered, too, that it is not possible to predict whether a pay contains only water with some traces of gas or oil or a real oil- or gasfield, because the mud gas show may be nearly the same, or even tend to the wrong side.

6.7   Chromatographic Mud Gas Analysis

The best and most complete way of mud gas analysis is made by gas chromatographs. These detectors suck a small portion of gas out of the gas stream, wash it with helium into a molecular sieve, and wash it out of the sieve into a detector of the cold wire, hot wire or F.I.D. type. The molecular sieve consists of a copper tube with a filling of charcoal, brick dust or some other material with a very high active surface. The different types of gas are separated by a molecular sieve, and enter the final detector one by one, the smallest molecules first. The cross potential of the end detector shows a lot of peaks on the chart, each corresponding to one type of gas. So it is possible to make an entire gas analysis within less than ten minutes, and a mud gas log for the entire well or at least for the time when the well is drilled through interesting regions. Of course such a detector needs to be controlled by hand though it works automatically in principle, and mud gas logging may become a very expensive part of geological work. On the other hand it may become much more expensive if a well blows out, or if a possible pay has been overlooked.

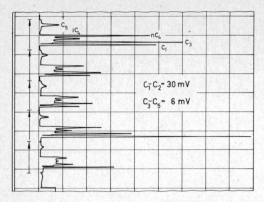

Fig. 78 Continous diagram of a gas survey by a gas chromatograph. The
quantity of each type of gas can be traced very easily and exactly.

6.8   Mud Gas Logging and Blow Out Prevention

A well site geologist should never forget that the first gas
logging units were invented for blow out prevention only, and that
the recent use of such units for geological purposes should not be
at the expense of blow out prevention. Such blow outs may occur
in general when there is too little mud weight within the well, either
because the mud has too little density, of because the mud column
has dropped as a result of severe fluid loss.

Mud gas detectors in general are only very bad warning instruments,
because they say nothing about the real gas contents within the mud,
and nothing about the mud level in the well and the overflow. The
percentage of gas which is recorded on some chart says nothing
about the percentage of gas contained in the mud, it doesn't give
a true picture of the mixture of gas and mud at the gas trap. Gas
recorders with alarm bells and similar facilities may be very effective
in practice, but in principle they are not to be trusted. They sound
the alarm even if there is really no reason to fear a blow out, or there
may be a blow out while the recorder shows zero.

The reasons for this lack of efficiency are obvious. The only facts
of real interest for blow out prevention are the density of the mud,
and the mud level in the well. Modern gas logging units therefore
comprise a mud level recorder, and a density recorder. These units
are comparatively inexpensive, and they may really give warning of
blow outs. In dangerous regions there should be a direct connection
from these units to the dog house, and a chart which informs the
driller how his mud density and level are at the moment.

6.9 Logging the Density of the Drilling-Fluid

The only real possibility of recording the density of a fluid like drilling-mud while it is flowing through some channel or similar line is by nuclear methods, by gamma-gamma-(density) logging, or by neutron-neutron-(moisture) logging.

Gamma-gamma logging is used for a number of different puposes. It is one of the most convenient methods used for density measurements. The analyser contains a source of cobalt-60 or caesium-137 and a scintillation counter. Of course it has to be entirely enclosed according to the laws concerning fire hasards on oil wells.

The principle is very simple and generally known. The source emits gamma rays, which easily pentrate even through the walls of the mud channel. Part of them will be scattered back to the outside, and this rate may be counted by a scintillation detector. The counting rate of the detector corresponds to the density of the material from where the gamma rays are back scattered, in this case by a constant background and the drilling mud. If the detector is put on the same float as the gas stirrer, is may be in direct contact with the mud, and the background can be minimised.

If the mud level sinks downhole however, and there is only air within the reach of the detector, it will show zero density, and that would mean immediate alarm.

The gamma sources which are necessary for such a density analyser may be very weak, and without danger to the drilling and gas logging personnel. Logging the water contents of the drilling mud by neutron/neutron methods will lead to similar results. For this method a source of Americium/Beryllium should be used. The counting rate corresponds to the hydrogen contents within reach of the analyser, and so to the water contents.

Neutrons may penetrate steel plates without loss of energy. So this method should be used if the sensor has to be applied to the outside of the casing head or to the outside of a mud channel, while the gamma/gamma-method should be preferable to all direct applications, such as on the gas trap.

There are some special mud level recorders available, which are usually kept floating on the mud surface within the mud funnel. They work on different principles. Most contain some potentiometer or similar electrical arrangement which records the height of the float within the funnel, or the angle of the arm which keeps the float in place. Others contain only a perpendicular tube with a pressure recorder on its upper end, which logs the level of the mud, or similar arrays.

In general, such level recorders may become superfluous if the density recorder is applied in the right way to the mud surface, or a neutron detector is fitted to the outer side of the casing head.

6.10   Gas "Tails" and False Gas Alarm

Most drilling muds are so „stiff", they have such a high gel strength and such a high viscosity, that a big part of the minor gas bubbles will remain within the mud when it has passed through the shale shaker, and remain in it until it reaches the suction tank. Then this mud will be pumped into the well for a second time, the same gas will appear in the gas log like a second, but smaller peak, and perhaps even for a third time. Such „tails" of course are only clearly to be seen on the gas record if the pay which produced the gas is not too thick.

A well site geologist therefore should know the time that is necessary for one round trip of the drilling mud. This time depends on the output of the slush pumps, the depth and the calibre of the well, and of the size of the drilling string. Drillers usually have a special slide rule for this calculation, but in general it can be calculated easily by hand, if the output of the slush pumps is known.

If the drilling has been stopped for some time, and recommences its work, there are often false gas alarms caused by small quantities of gas which collected on the well head during the dead time. Such false alarms usually occur on Monday morning, when the drilling has been stopped over Sunday. Of course it would be nonsense to close the preventers in such cases, and a short look at the recorder will always show that the percentage is diminishing rapidly, and after an hour or so the recorder will show zero or only traces. A good contact between driller and geologist will save precious time in such cases, but on the other hand there is no law that forbids blow outs on Mondays, and the recorder should be checked immediately if the gas logging unit shows the red light or the alarm bell is ringing.

If only simple cold wire detectors are used for the logging unit, many false alarms may occur if sour or corrosive drilling-mud is used. Such types of drilling mud react with the iron of the drilling string, and hydrogen will develop. This type of detectors is more responsive to this gas than methane or other hydrocarbon gases, and a gas alarm may be even produced by small traces of hydrogen.

In such cases another detector should be used, or another type of drilling mud. Original hydrogen may occur too in a well, but this gas is extremely rare, and some specialists even say that reports on natural hydrogen in such wells are erroneous or simply not true.

6.11   How to Analyse the Gas Contents of Cuttings

The cuttings coming from the shale shaker are contained in empty tins or plastic pots and are subjected to a series of tests which are made by samplers in routine work.

Cuttings often contain gas bubbles, and sometimes it is desirable to know what kind of gas it is, and if it is not the „tail" of another pay some hundred metres higher up. To make an analysis, the mud sample is put into a mixer without washing it beforehand, and the freed gas is pumped into a detector. The exhaust of the detector is reconnected to the mixer, or some fresh air may be added to the sample by a washing bottle or a valve.

Other mud specialists usually put a sample under higher vacuum pressure without mixing or stirring, and others usually heat the sample in order to free the gas. Experience, however, has shown that heating may lead to erroneous results, and that vacuum alone will not give good results. So the above mentioned procedure seems to be the best.

The type of detector used for this kind of analysis depends on the type of gas in that region and that pay, and on the results of regular gas logging. Of course the gas may be fed into a chromatograph or an infra-red detector if it seems necessary, and if such facilities are available on the well site.

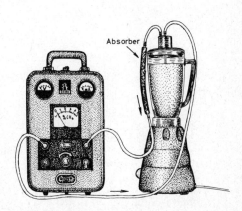

Absorber

Fig. 79 How to detect gas traces in cuttings. The sample is fed into an electric mixer. The gas traces are sucked off via an absorber with silicagel and the percentage of the gas contents may be read from an usual hot-wire detector or a similar gas detector (Courtes. Geomessdienst).

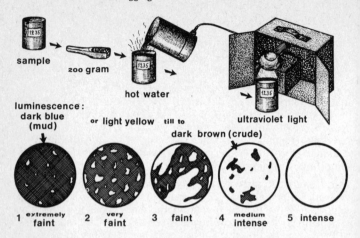

sample

200 gram

hot water

luminescence:
dark blue
(mud)   or light yellow till to   ultraviolet light
         dark brown (crude)

| 1 extremely faint | 2 very faint | 3 faint | 4 medium intense | 5 intense |

Fig. 80 How to detect oil traces in cuttings. The sample is fed into an empty tin, and the tin is filled up with boiling water. All droplets of oil will collect at the surface of the water, where their amount will be controlled under ultraviolet light (after Schettler).

6.12 How to Analyse the Oil Contents of Cuttings

If the drilling mud is very stiff, the tin or the pot containing the cuttings and the gas bubbles may be carried over longer distances to the next laboratory to make a thorough analysis there with better facilities.

Hot Water Test

At first a quantitiy of the sample is put into an empty tin, hot water is poured over it, the mixture is stirred, and after some time it is examined in the dark by a quartz lamp. Any droplets of oil contained within the mud or in brickles of rock will come to the surface, and will show up as yellow, brownish or dark brown spots. This luminescence is specific to oils, and crude oils usually show light yellow to dark brown luminescence. White or yellowishwhite colours are rare with crude oils. Lubricating oil, Kerosene and other types of hydrocarbons which have passed through a refining plant show greenish, bluish, reddish or similar colours, but never yellow or brown. So it is easy to see whether a droplet of oil is real crude, or whether it comes from the engines or from the lubrication of the drill string.

The quantity of crude covering the surface of the fluid within the tin may be guessed, and a scale of one to ten or one to five may be noted for every sample.

Chloroform Test

If a sample shows crude on the surface after the hot water test, a bit of the sample is rinsed carefully on a sieve of one or two millimetres, and the residue remaining on the screen is checked under the quartz lamp, or under a microscope with a magnification of 1:15 or 1:20. All brickles with apparent luminescence or crude are picked out with a pair of tweezers, and put into a clean mortar. There they are crushed and put into a folded paper filter, where some chloroform is poured over it. After some time the filter is checked under the quartz lamp. Any oil contained in the brickles will have been dissolved by the chloroform, and the solution will creep up the filterpaper till the solvent evaporates. Crude or other oil contained in the solution will form a shining ring where the solvent evaporates, and the colour of luminescence and the quantity of brickles should be noted.

The rest of the solvent with the dissolved oil is put into a test tube, and another test tube is filled with some cubic centimetres of the clear solvent. Some water is added to both tubes, and then they are shaken for a while and compared in front of a black sheet of paper or in transmitting light.

The test tube containing dissolved oil will appear a bit whitish, because oil has separated from the solvent and forms a kind of emulsion with the water. Chloroform and similar solvents show a better affinity for water than for oil, and if both of these are present, as in this case, oil will get out of the solution.

Such oil show is very important for the geological report, and all grades and stains of luminescence, and cuttings with traces of crude have to be noted in the so-called Sampler Log.

7 Cuttings and how to Analyse them Stratigraphically

7.1 Petrographic Analysis of Cuttings

One part of the sample is washed on a sieve of one millimetre and dried on an electric heater, or below an infra-red lamp. The dry sample, which has to be completely free of mud and has to show its natural stains, is inspected under a microscope with a magnification of 1:15 or 1:30. This is done on a tray with a grid so that rough percentages may be guessed. At first brickles of the same type of rock are picked out of the sample and arranged in one corner of the tray.

Brickles of limestone usually are angular, they react very quickly and very actively when treated with a 1:4 solution of hydrochloric acid, and will float up to the surface if they are thrown into a small receptacle with that acid.

Brickles of dolomite may show similar shape and form. They will not react with cold hydrochloric acid, or only along some corners. If a brickle is crushed in a mortar, however, it will react very quickly with that acid, and the powder will float on the surface of the acid between the gas bubbles.

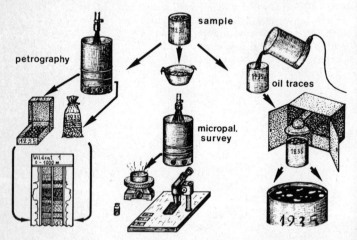

Fig. 81   Some of the different methods of testing cuttings. Part of the sample is washed and dried for petrographical analysis. Another part is digested, washed through a set of sieves and dried for micropaleontologic work. Another part is tested for oil traces by the hot water method.

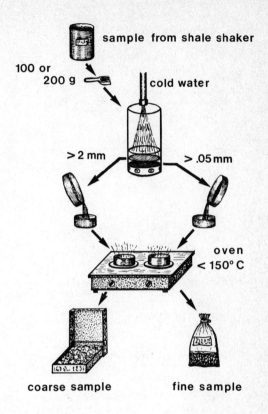

Fig. 82 The preparation of cuttings for petrographical analysis. A certain quantity of the sample is washed through a set of sieves, and the two fractions are dried for closer inspection.

Limy marls usually show rounder forms than limestone, and they too float up to the surface in a receptacle with acid. After the reaction, however, a thick smear of mud will remain on the bottom.

Marls with less lime generally move around in the acid pot, but they will not float up to the surface because the reaction is too weak. Shale, claystone and mudstone will not show any reaction with hydrochloric acid. They may be scraped with the steel needle very easily, and even with the thumb nail or the spine of a hedgehog. Gypsum and anhydrite sometimes look like calcite or limestone, but they are very soft, they may be scraped easily, and of course they do not show any reaction with hydrochloric or other acid. Quartz and quartzite may occur in every shape or form, from angular brickles to well rounded pebbles and sand grains. They show no reaction with any

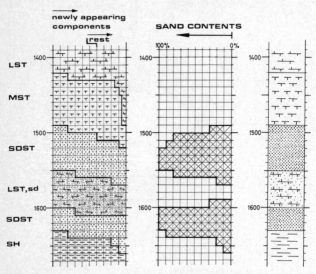

Fig. 83 How to present the results of petrographical inspection of cuttings.

kind of acid, and they cannot be scraped by the neddle, but he needle even leaves a tiny metallic or black trace on them. They will scrape a glass plate when they are moved over it, and in general they are very easy to detect. Beginners usually have some difficulty with mixtures like sandy claystones or clayey sandstones. In reality these differences are not very pronounced in rough routine work. The easiest way to name such mixtures is to look at them and to see, what part is most striking. So if a brickle looks like a sandstone, but shows some mud particles under the microscope, it should be called a clayey sandstone. If it looks like a claystone, but shows some fine or coarse sandgrains under the microscope or when ground between the teeth, is should be called sandy claystone. Sands and sandstones are most important pays, and all sand and sandy brickles should be tested with the utmost care.

Brickles of sandstone are checked under the microscope in total, then they are crushed and inspected once again. Finally they are tested with hydrochloric acid in order to see if they have calcitic, clayey or other bonding.

The average size of the sandgrains is measured on a sheet of gridpaper, or better with a so-called sand grain lens. Finally all types of different petrographic units in one sample are listed, beginning with the most abundant and ending with that showing the smallest quantity. Their percentage is guessed from the number of centimetres they cover on the grid tray, and finally they are noted on the sampler log.

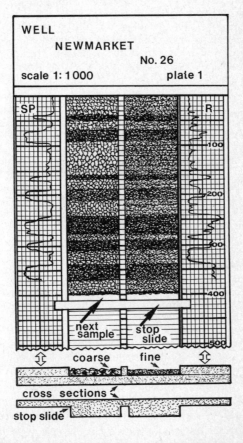

Fig. 84 Section of the petrolog of a well, with one strip of coarse cuttings and one of fine cuttings glued on a plate of plywood, together with some bore hole logs.

The sand contents of every sample is weighted on a balance, or measured in a cylinder with a grid outside. The sand contents are noted in a special space on the sampler log. In regions with terrestric layers or changing petrography it may be very useful to stick part of the washed and dried cuttings on a wooden plate together with some bore hole logs, in order to get a good and quick survey of the petrography, and especially of the changing natural stains.

The plate of this plywood or similar material contains a flat channel where the cuttings are fixed by a normal plastic glue. For every five or ten metres a corresponding small strip of that channel is covered with glue, the washed, dried and sieved cuttings are poured on the glue and allowed to dry. After some minutes the surplus cuttings are brushed off, and the next strip may be added. For routine work the lower part of the channel is blocked by a fitting piece of wood, and only the stripe which is going to be worked is left open.

Care should be tanken, however, to ensure good ventilation in the well site laboratory, because all solvents may cause headache or even serious trouble to the liver or to the intestinal nerves, if their concentration remains too high during a lengthy spell. When the well is finished, and all cuttings are glued to the panel, the entire strip should be sprayed over using some diluted glue of similar type, or a plastic vernice, which is not too shining or too bright. If the right type of vernice is used, the natural colours will clearly show up, and terrestric parts with red, brown or green stains will be easily detectable within the grey, bluish, black or white lacustrine or marine layers.

7.2 Biochronological Analysis of Cuttings

Cuttings usually contain not only material from the formation which has just been drilled through, but some backfall of higher horizons too, and they are never from the depth where the bit was rolling when the sample was taken, but from some horizon several metres higher up. The value of cuttings for stratigraphic dating, therefore, is less than that of a core, but they are much cheaper to obtain, and with some care they may deliver first class results.

There are several possibilities of making mistakes and getting erraneous datings, but most mistakes are caused by lack of care only. A well site geologist therefore should watch and train his samplers

properly, and see to avoid mistakes during the analysis of the cuttings. At wildcat drilling or offset wells, samples are taken usually every two metres, or every five metres in production wells, which are not so interesting for stratigraphy.

When the sample has to be taken, the driller usually rings a bell or a claxon. The sample should not be taken immediately, however, but only after cleaning the shale shakers from previous cuttings. Then the pot or tin is filled with a real fresh sample, it is labelled accurately, and brought to the mud lab. Usually the different types of analysis are not made one by one, but in series. Only if a special horizon or pay has to be cored, the samples have to be worked one by one, and perhaps with reduced drilling speed.

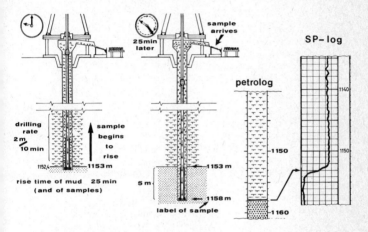

Fig. 85 How to date the depth of a sample correctly. The "label" depth of the sample is some meters deeper than its real depth. For the correction you need the size of the well, the efficiency of the mud pumps, and the rate of penetration.

If some mud sample or gas peak has to be compared with a bore hole log later on, the depth of the sample must be dated correctly. This may be done calculating the „rise time" of the mud versus the penetration rate.

The rise time of the mud is the time necessary to move from the bottom of the well up to the shale shaker. It may be calculated by the special slide rule of the driller, taking into account the output of the slush pumps, the size and depth of the hole, and the size of the drilling string.

The penetration rate may be taken from the drill recorder or a similar recorder. Thus if the rise time is 45 minutes, and the penetration rate is 9 metres/hour, the sample will be from a horizon 6 metres higher than it is labelled.

In general such corrections are not necessary for smaller wells, especially if the final completion logs are represented on a scale 1:1000. For special purposes and precision work, however, it will be necessary to redate the samples and to correct the samples from their apparent depth to their real depth.

7.3   How to Extract Microfossils from Cuttings

A mud sample in its raw state may contain a mixture of several different formations, but only fossils from the lowest formation are of real interest. A big part of previous formations may be washed away with the fine part of the mud on a 2 mm screen, leaving only clean cuttings on the coarse screen. It is a matter of experience if hot water, warm water or only cold water may be used best for washing. Some young, poorly consolidated formations may completely disappear through the sieve, if hot water is used for washing.

The cuttings should be dried then at least superficially, and then they are put into a halfround pot of aluminium or stainless steel. Some 15 percent hydrogen peroxide is then poured over the sample until it is completely soaked. This chemical is sold in a 40 percent solution for chemical and technical purposes. It contains some weak organic acid for stabilisation, which does not do any harm but avoids explosions and disintegration during storage. After some minutes the sample will begin to show some movement, slight steaming and a kind of soft cooking. This shows that the chemical is going to disintegrate within the pores of the cuttings. The bubbles of oxygen arising from this chemical reaction will destroy the texture of most kinds of rock, with the exception of hard limestones, dolomites, consolidated sandstones and similar hard rock. The reaction will usually finish after some twenty minutes. If it seems to take longer than usual, some drops of potash or carbonate of soda will speed it up, because the stabilisation by the organic acid will then be broken.

On the other hand a very quick and sudden reaction may create very hot temperatures with some brown discolouring of the sample. That should be avoided because all pyritised fossils may then be destroyed by reaction with the fresh oxygen.

As soon as the reaction has ceased, the sample is washed through a fine sieve, so that the mud is washed away and only the microfossils, sand grains and similar things are left. This residue is dried on an oven, on an electric cooking plate, or below an infra-red lamp. Then it is put into a small glass tube with a plug or cap, and with a correct and complete label. The microsamples are checked under a binocular microscope with medium or high magnification, and a complete micropaleontological analysis can be made of the contents. If necessary, the forams or ostracs or conodonts are picked out of the sample by a needle treated with bees wax, and put into a special cell with a removable transparent cover.

There are two principal kinds of error possible when such faunules are used for dating formations in a well. First of all the sample may contain a mixture of the formation where the sample comes from, and of younger backfall. That cannot be avoided completely, even with the greatest care, but backfall may be reduced by good mud practice, and by exact sample-taking practice. An experienced sampler will soon know what part of the faunule really belongs to the formation which has been drilled at that depth, and what part of the fossils comes from the backfall.

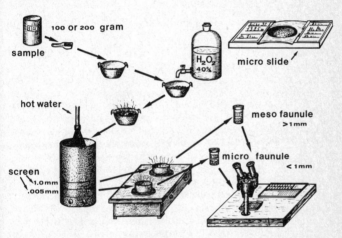

Fig. 86 How to detect microfossils in cuttings. A certain quantity of the sample is digested by a 15 % solution of hydrogen peroxide. The mud resulting from that treatment is washed through a set of sieves, and the residues on the screens are dried. The coarser fraction may contain mesofossils. The finer fraction has to be inspected under the microscope. The microfossils are picked into special micro slides.

The second kind of error is much worse and may even lead an experienced worker to false datings. If a formation without any microfossils or with only very scarce fossils is just drilled through, and a formation with abundant microfossils is lying above it, the samples will often contain a rich, but false faunule, and no microfossils of the real formation. It is only a matter of regional experience to know such facts, and to date the real formation by some other means, such as the colour of the cuttings, the petrography or its different rate of penetration.

7.4   Acetate Films Made from Cuttings of Hard Formation

Cuttings of hard formations, like limestones, dolomites, hard sandstones and similar rock, usually cannot be distintegrated by hydrogen
peroxide, or even by other methods which take more time, like
cooking with sodium sulphate, or continous thawing and freezing.
The easiest method of stratigraphic analysis in such series is to make
thin sections of some cuttings, and to look them through under the
microscope with transmittied light. This method, however would
consume too much time and cannot be applied in practice. There are
two methods which take much less time, and which in general show
very good results: the method using acetate copies of cuttings, and
the method using insoluble residues.

Acetate copies need only little time, if the mud lab is supplied with
a grinding machine.

The cuttings are dried and are put into a ring made from a tube of
plastic, which is stuck to some cardboard or some thin folio. Then
some epoxy resin is mixed with hardener according to the usual
recipe, poored on the sample, and the resin is allowed to harden
thoroughly. Then the ring with its contents is ground on the machine,
with coarse emery first, and finally with 500 mesh powder. The
polished surface is etched then for some seconds in dilute monochloracetic acid, or titriplex, or formic acid, or some other organic
acid which attacks limestone. Acetic acid works only very slowly,
formic is much better in general. The percentage of dilution, and the
etching time must be tried for every type of limestone. After rinsing
and drying, the ground surface is covered with acetone, and a sheet of
acetate film is rolled on the surface. Some slab of marble or a heavy
glassplate is put on it, and the acetone is allowed to evaporate. The
acetate film is then drawn off, and put between two glass plates for
examination under the microscope. For the determination of microfossils it is best to use a so-called dark – field – condenser, or to use at
least oblique light. The images of microfossils printed on the acetate
film may be photographed or treated like any thin section. If there
are any forams or other fossils contained within one slab which show
a very complicated texture, it is possible to make several copies from
one object after subsequent grinding and etching, and to make a
reconstruction of the fossil by a series of drawing or photographs.

This method sounds very complicated and time-consuming, but in
reality its results show up very quickly, if a sampler is used to do that
work, and does it in series. The films may be labelled and stored like
all other material.

For higher magnifications it is often desirable to produce a
preparate with better optical qualities than these films. It is not
possible, however, to trim a small piece of film out, and to prepare
it with canada-resin and a glass cover like a histological section,

because the images of the fossils would then disappear. For such purposes a resin with a higher or lower refractive index should be used. Good results have been obtained with a thick solution of Aroclor in toluene, with a refractive index of 1.66, for instance.

7.5 Insoluble Residues Made from Cuttings

Insoluble residues may be sufficient in some regions to identify a limestone or a dolomite without too much work. Some cuttings, usually about the same quantity, are put into a test tube and etched with dilute hydrochloric acid, monochloracetic acid, or formic acid. Regional experience has to show which acid works best, and what dilution should be used.

After the end of the reaction, the test tube is shaken a bit, it is filled completely with water to the top, shaken again, and after some seconds, the water is decanted with some care, and the rest is washed with some water on a filter. The water is then washed away with some cheap kind of alcohol, such a methyl alcohol, and the filter is then dried and its residue is put into a small glass pot, or into a cell for microfossils, or it is mixed with canada resin and smeared on a glass slide, covered by a glass plate, and used like any usual microscopic preparate.

The types and frequency of these insoluble materials are determined under the microscope, they are listed and put on record like fossils or other materials used for stratigraphy. In some regions this method is the only one which may be used, and the results are sometimes surprisingly exact.

7.6 Dating Cuttings by Nannofossils

Since some time it has been made possible to date quarternary, tertiary and even cretaceous formations by nannofossils, mostly calcareous remains of coccolithic algae. These minute shells are much smaller than usual microfossils („mikro" in greek means small, „nanno" means dwarf) and they may be seen only in a good microscope with a magnification of 1:1000 or more, or better in a scanning electronic microscope. The preparation is very simple, if the material is not too hard or too much consolidated.

A small chip of a cutting, not bigger than a fingernail, is put into a test tube, and some drops of hydrogen peroxide are poured over it. After the end of the reaction, the test tube is put into an ultrasonic shaker with a frequency of 20 or 100 kHz, and the contents are vibrated thoroughly for five or ten minutes. Then the mixture is allowed to settle, and when the overlying water has become clear, a very small drop of the uppermost surface of the sediment is taken out by a pipette with a long, fine tail, and that drop is spread on a glass plate and allowed to dry. Then the dry surface is covered with a very fine vernice which is sold in shops for drawing utensils under the name of fixative spray. After drying, the preparate is ready for use. It may be covered with a drop of immersion oil and looked through with an oil immersion lens, or it may be let dry if a lens for dry magnification is available. Most of these minute tests are so small, that even very good lenses show only part of their structure, and magnifications of 1:10,000 or more have to be used to get all the details. In general, however, a good microscope with a good immersion lens will be sufficient for routine work.

It goes without saying, however, that it needs some time of training, some years of experience and a good morphologie memory to work with that material, as for all palaeontological work. Anyhow this method enables a good operator to determine even a small chip of one cutting, and to determine it very exactly and thoroughly. One of the biggest advantages of this method is the fact that even sands and poorly consolidated sandstones, which usually contain no micro-fossils, may be dated by such nannofossils which are contained in their pore space or in their material, and which can be prepared by the same simple procedure.

8 Cores and Core Analysis

Drilling cores is a very expensive procedure, it may cause deviations of the bore hole, and even blow outs. In former times it was usual to core entire wells from the surface to the bottom, and even nowadays smaller wells may be cored because the slush pumps are too small or because the rock is too hard for usual rock bits.

In general, however, cores are restricted to the pay zone, and every metre of coring may cause long discussions with financial surveyors and partner companies.

It needs some skill to place a core properly into the top of the pay zone, and to get cores of all porous and permeable parts of the pay.

Modern core barrels with diamond crowns, free running interior barrel and a length of more than twenty metres enable a good driller with appropriate experience to drill some twenty metres of core without losing important parts of the pay, and without losses at the crown.

Anyhow a well site geologist should know as much as possible about cores, core losses, and core drilling in general.

8.1 How to Place a Core properly

A newcomer would think that it is completely impossible to drill a well down to more than two or three thousand metres, and to place a core directly into the top a of a pay, though the next well is some miles away, and geophysics rarely predict more exactly than fifty or sixty metres.

In reality it is possible to drill cores so exactly, and a well site geologist has to learn methods which enable him to core exactly and to avoid gaps in the pay zone as well as useless cores.

The most important basis for good coring is a constant, exact stratigraphic survey of the entire well, with thorough dating of all cuttings, and constant gas logging.

Sometimes there are zones above the pay with characteristic microfossils at a permanent, equal distance from the head of the pay. Then the cuttings in that part of the hole should be taken not only every 5 or 2 metres, but every single metre. The pumping rate should be increased, the penetration rate should be reduced, and the critical zone with the special microfossils has to be calculated as exactly as possible. Regional experience has to show if the distance between marker horizon and pay may be trusted really, or if there is some shifting in this or that direction.

One of the best methods to place a core exactly may be a bore hole log, compared with a similar bore hole log of a neighbouring well.

Sometimes a string of casing is run between the gas anchor and the lower part of the well, and of course a bore hole log is run before running the casing. If the thickness of the layers between the end of the log and the pay is very stable, the log may be correlated with one or several logs of wells in the surroundings, and the distance to the head of the pay may be calculated.

Some good indications may come from gas logging. Usually every stratigraphic formations has some characteristics as to the gas logs, some streaks of sand with residual oil or methane dissolved in water or similar special features. If the gas logs from one well are compared with those of other wells in the surroundings, it will always be possible to find some parallel trends, and to compare peaks and ratios with one another. Traces of residual oil in some sandy marl, which are devoid of any commerical interest, may become most important for placing cores properly when they appear in gas logs, when they can be correlated from one well to another, and if they appear at a constant distance above the top of the pay.

Besides, nearly all pays which yield oil or gas, or which have yielded such hydrocarbons in earlier geological times show a kind of gas aureole above their top, if gas logging is done with good equipment and very sensitive gas detectors. Such pre-warning may begin more than fifty metres above the pay, and in some cases even higher. As with all other methods these gas shows have to be compared with gas logs in the neighbourhood, and the density of the gas shows seems to depend on the type of mud and other features, too. Equal gas shows can be expected only if the same type of mud with equal density has been used, and if the rate of penetration is more or less equal. Even the type of bit may play some role, the pumping rate and other special features.

8.2   Direct Methods of Coring a Pay

Even the best method may fail one day, the calculated distance may be wrong, or a fault may have interrupted the series, or some disconformity may cause difficulties. Even in such cases the pay need not be rolled through, if driller, samplers and geologist keep their eyes open and work together.

Pays usually show a different penetration rate when compared with the overlying formation. Sands usually show a better penetration rate, sandstones and limestones usually a smaller one. So the driller should watch the drill recorder exactly, and if there is some change, he should stop penetration and wash the well through, to get the samples and the gas bubbles to the surface. If the cuttings really contain parts of the pay, and if the gas detector shows some peak of gas or oil, the drilling string should be hauled up, and a core should be drilled.

In wildcats far from every other well with which they could be compared, such direct methods are sometimes the only possible procedure. They need constant alertness, however, and good nerves on the part of the drilling and the geological personnel.

In practice it is not a real disaster if a sand, sandstone or some other pay has been drilled through without coring. The important features of every pay may be determined later on by side wall coring, bore hole logging and testing. Experience has shown, however, that a core cannot really be replaced, and that only all the possible methods, including core analysis in a petrological lab, will lead to a real understanding of the structure and to understand its properties in every respect.

8.3   How to Drill a Core

Core drilling is expensive because at first the entire drilling string has to be hauled up, unscrewed and put aside. Then the core barrel

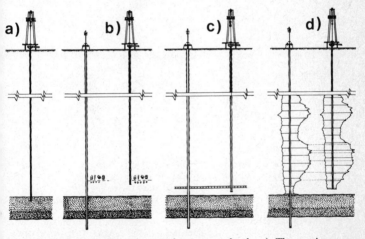

Fig. 87 How to begin coring at the correct depth. a) The cuttings are controlled very quickly and very exactly. As soon as brickles of the pay formation appear, the drilling string is hauled up and the core barrel is run in. Anyhow the first meters of the pay will get lost. b) Micropaleontological comparison with a nearby well. The cuttings and their microfossils are inspected very quickly and very carefully. Some special horizon some meters above the pay may enable a good sampler to begin coring just at the top of the pay. c) Petrographical comparison with a nearby well. Some special petrographical components some meters above the pay may be used in a similar way. d) If there are no possibilities to date the top of the pay by microfossils or by petrology, a bore hole log should be run. The top of the pay may be calculated after correlation with the log of a nearby well.

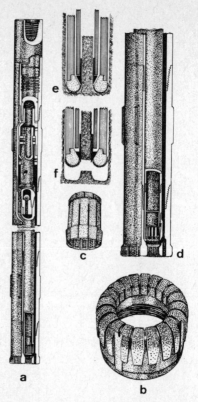

Fig. 88 A modern core barrel, and how it works. a) Cutaway view of the core barrel with its inner tube, its outer tube, and the diamond crown (b). c) The core catcher, a very sophisticated type of steel spring. d) The lower end of the core barrel with the core catcher, the crown, and the conus which makes the catcher grip the core as soon as the barrel is hauled up. e) Section across the barrel during drilling. The core is cut out of the formation by the crown and slips into the inner barrel through the catcher. f) If the barrel is hauled, the catcher is pressed together by the conus. The core breaks off and remains within the core barrel.

has to be screwed on the lower end of the string instead of the bit, and the whole string has to be screwed together length by length, and run into the hole again. Core barrels are relatively tiny in comparison with rotary bits, and drilling has to be done slowly and with only little weight on the bottom.

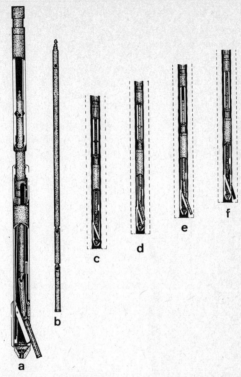

Fig. 89 Side wall coring. The entire core barrel fits into the annulus of drill pipes (a). Only the lowermost part may take up a core of about one inch in diameter. The core barrel is lowered down by the sand line, a core may be drilled out of the bore hole wall, and then the barrel is retracted and hauled to the surface (c–f).

There a numerous types and sizes of core barrels. The oldest type consists of one tube only, with teeth of hard metal on the lower margin, and a kind of spring which catches the core just above these teeth. These simple core barrels may work perfectly in hard formation, but will not be any good in a nonconsolidated sandstone or a soft marl or mudstone. Nonconsolidated sandstones will be washed away with some certainty, and from a pay consisting of hard streaks without oil and porous sandstone with oil there will remain only the hard streaks, giving a completely false impression of the oil potential.

For softer formations there are core barrels with a double wall, which avoid the mud stream to wash the core out of the barrel. Older types of such core barrels have a solid connection between the two tubes, and tend to twist soft marls to a kind of smear with a very complicated structure. These types meantime have been replaced by so-called „free running, double core barrels", where only the outer tube is connected to the drill string and bears the teeth or diamond crown for cutting the formation. The inner tube is running on roller bearings and may remain standing still, while the outer tube turns around and drills the core.

The lower margin of such core barrels is equipped with teeth of tungsten carbide, with small rollers or usually with a crown of diamonds which cut a ring into the formation. The part of rock within

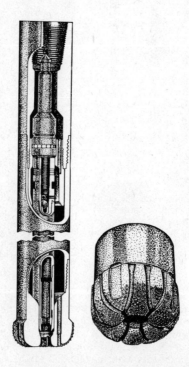

Fig. 90 Wire line coring. The core barrel fits into the annulus of the drilling string and enables to core the formation more or less continously, but at a diameter of about one inch only. The barrel may be hauled and inserted by the sand line.

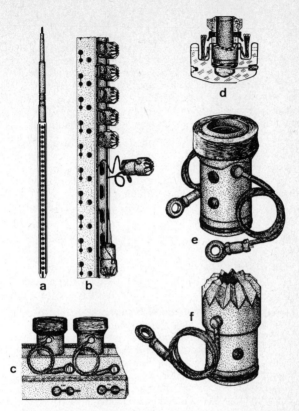

Fig. 91 Shooting cores out of the bore hole wall. The core barrel (a) is lowered to the desired formation using the logging cable. The barrel contains a certain quantity of very small steel tubes, which are connected to the barrel by steel ropes (b). Below each tube there is a small explosive charge which may be fired by the logging cable (d). Core tubes for soft formation have a wider opening than those for hard formation (e, f).

the ring will slip into the core barrel when drilling goes on, and fill the inner tube of the barrel until drilling is finished. Then the drilling string together with the core barrel is hauled up. As soon as the barrel moves upward, a special ring supplied with a cone will grip around the core, will keep it fast and tear it off the bottom of the hole.

Of course the ring, called the core catcher, may slip a bit, or some very soft part of a sandstone may be washed away despite all precautions. In general, however, such „free running double core barrels" are very reliable in relation to the older types, and often show hundred per cent core recovery, while older types often will bring no recovery or only a modest percentage. Such modern barrels are sold, rented or used in service in nearly every length beginning from about three metres up to about twenty-one metres. Such long core barrels are relatively delicate instruments, and many drillers rent them from a service company together with an engineer who watches the handling, the rotating speed, the load and the hauling manoeuvre.

Before another core may be drilled, the hole usually has to be reamed, which means another round trip in and out with a rotary bit instead of the core barrel. Such round trips cost a lot of time, even if he hole is not too deep, and therefore cores are very expensive, and core drilling has to be reduced to a minimum.

8.4   Side Wall Coring on the Logging Cable

If a pay or a possible pay has been drilled through, or a new sand or limestone shows up in a bore hole log, there is still a possibility of coring it, even if the well is some hundred metres deeper meanwhile.

The most common method for such purposes is the so-called side wall coring gun, used by logging companies on their logging cable. It is a heavy steel bar containing about twenty small core barrels of about one inch in diametre. These barrels contain an explosive driving charge and are fixed to the barrel by small steel ropes. Every single core barrel may be fired into the wall of the bore hole by press button operation, and when the barrel is drawn up, it draws the barrel out of the formation, and the next barrel may be fired.

Of course several barrels may be shot without moving the gun upward, and the different core barrels may be distributed over the entire well if that seems necessary or desirable.

This method seems to be very efficient at first sight, but in reality there are some disadvantages which make the results of this kind of side wall coring less efficient.

The most important disadvantage is the small size of the core. There are two types of such core barrels, those for soft and others for hard formation. The barrels for hard formation have an inner diameter of about 22 mm, and those for soft formation are a bit thicker. The maximum length of a core is about one inch or a bit more, and most times is it is even much shorter. On the other hand, the texture of a sandstone or limestone will be destroyed completely by the procedure of firing, and only brickless without real coherence will remain when the core is taken out of the barrel by the logging personnel. Another disadvantage is that these brickles come from the real bore hole wall, which is flushed by the drilling mud and shows completely altered pore contents.

So such cores shot out of oil bearing pays will show nearly no oil and have to be looked through under an untraviolet lamp for oil traces. Cores from sands with residual oil only may look nearly alike, and gas bearing fromations usually will show no gas traces at all, or gas traces which could come from the explosion. Porosity measurements on such brickles with destroyed texture are of problematic value, and the permeability cannot be measured with certainty, either. The only analysis which really can be made is a grain size analysis. With some uncertainty it is possible to calculate the porosity from a good grain size analysis, but permeability may show severe divergences from the norm, when calculated that way.

So side wall cores shot out of the bore hole wall are of doubtful value compared with a real core, but they are better than no core at all.
For technical and financial reasons such side wall cores should be shot only in accordance with a regular bore hole logging program. That will not only save some money, – and side wall coring is not much cheaper than real coring – but it avoids some trouble with several probes which have to contact the bore hole wall and could be disturbed or at least hampered by fragments of cores or core barrels.

8.5   How to Describe and to Analyse a Core

When the core barrel comes out of the hole, it is suspended a metre or so over the rotary table, then the opening of the table is closed by a steel plate, the crown is unscrewed, and the core catcher with the lowest part of the core comes out. The core barrel should then be kept only some twenty of thirty centimetres above the steel plate, and every piece of the core coming out of the barrel should be caught by the core tongue, put aside and lined up in core containers.

These core containers are about one metre long, they should be labelled before the core comes out of the hole, and strict orderliness should be observed. It is terrible work, even for an experienced geologist, to put a core into the right order which has fallen out of the core barrel without any control, or which has been put into containers without appropriate labelling, or even with a false label.

The entire series of the core is then carried to the well site lab, it is washed with cold water and set ready for analysis and description.

Any kind of analysis and description should be made immediately because the pore contents may be altered quite quickly by the influence of the atmosphere. Part of the volatile components may evaporate, and even the natural colours may change and become very different.

The first inspection should be made immediately after washing the core. The most important tools for that inspection are a hammer, a pocket lens, an ultra-violet lamp, and a flask with hydrochloric acid.

The very first inspection should concern the outside of the core. A sandstone with a good oil content will rarely show any oil to the naked eye. Cores with conspicuous oil show and oil drops are rarely good. „A bleeding core does not produce." The crossection of the core should show a good impregnation of oil in its interior part, and only a small ring of mud filtrate on the outside. Cores with water and only residual oil use to show just the contrary: The oil is concentrated within an outer ring, and the interior contains almost only water. The quartz lamp will show that better and more precisely than the naked eye. If the entire core shows a yellow or brown luminescence, and only the outer ring shows the dark blue-violet colour of mud filtrate, the core will probably produce oil. If the inner part shows dark blue or violet, and only an outer ring shows yellow or brown, the core is undoubtedly watered.

If only very scarce yellow or whitish points appear under the quartz lamp, they should be treated with chloroform in a mortar to see if they are really oil, or if they come from some carbonate crystals which have been „tortured" by the drilling process.

Cores from gas-bearing pays usually show nothing at all, neither gas bubbles nor anything else. If the mud log has shown any gas in that zone, one slab of the core should be tested for gas, and if there should be any doubt, this gas test should be repeated with other parts of the same core. Such a gas test needs some preparation, and one of the most important things is to select the appropriate parts of a core.

Coarse sandstones and vuggy limestones lose most of their gas contents during hauling, and the little that remains will disappear very quickly. The best subjects for these tests are finegrained, porous sandstones or fine oolithic limestones. Tight parts may still yield gas or traces of hydrocarbons when all porous parts are barren. The method of gas testing depends on the facilities which are available. A good gas lab should contain a core tester. Such testers are filled with a slab of the core, the rest of the space is filled with water, and the cover is closed. Then low pressure is applied by the hand pump, and some gas or some mixture of air with gas will appear within the top of the tester. This gas sample may be conveyed into a small gas container, of fed directly into a gas detector.

If such a tester is not available, there are some other possibilities of getting a gas sample form the core.

For the pumping method you need some vacuum pump, either motordriven, or working by a jet of water.

The core is put into a suitable plastic tube with a removable cap, or into part of the inner tube of a car. One of the tube's ends is closed completely, the other side only during the test itself. The valve is removed from the valve holder, and a plastic or rubber tube suitable for low pressure serves as the outlet. A normal gas container is applied to the other end of the tube, a washing flask follows next, and then the arrangement is put under low pressure for only a short time. Then first the outer vent of the container, then the inner one are closed, and finally the pump is stopped. This first sample usually contains more air than gas, and should be rejected. The second sample is taken differently. First low pressure is applied to the gas container, then the outer valve is closed, and the inner one opened. This sample usually is better than the first one, and this procedure may be repeated several times.

The gas containers may then be fed into a detector, or sent to some lab with facilities for exact gas analysis.

Caution: Normal gas containers have fairly thin walls, and tend to cause implosions when low pressure is applied to them. Either metal containers, or containers from thick, transparent plastic, or solid glass containers should be used, and a wet towel should be put over the container to avoid imploding glass containers endangering the operator. Good success without danger will be achieved if champagne bottles are used for containers. Two holes are drilled through the original stopper, the two glass or plastic tubes are inserted as shown in the drawing, and the stopper is run into the neck of the bottle. Then another hole is drilled into the plastic stopper, and the empty space is filled with some epoxy resin mixed with hardener. These containers will not implode, even if a first class vacuum pump is used, and they need not be packed into special wooden chests for transport to prevent them from being broken.

If no core tester or vacuum pump is available, the gas may be driven out of the core by hot water. For that purpose the slab of core which is to be tested is put into an ordinary bucket, and boiling water or at least hot water is poured over it. A funnel of plastic or glass is kept over the core upside down, and the gas container is connected to the upper end of the funnel.

8.6 Testing the Properties of Sandstone Cores

The pay formation may be a sand, a sandstone, an oolithic limestone, a vugular limestone, a dolomite with primary or secondary porosity, or even a vulcanic or methamorphic type of rock. The most important properties of such a pay formation are its thickness, its porosity, and its permeability. These properties are very important for the quantity of oil or gas which can be contained within a

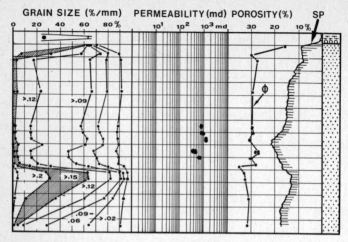

Fig. 92 Grain size analysis, permeability, porosity and self potential log of a sandstone in NW Germany (after Füchtbauer).

reservoir, the quantity which can be produced per day, and how it can be produced.

The thickness can be derived from any log with good resolution, e.g. a microlaterolog, a microlog, some other focussed array, or from the short normal. The thickness derived from a log, however, is only the apparent thickness, and most times it is somewhat greater than the true thickness, which must be measured perpendicular to the bedding plane of the formation. This reduction from apparent thickness to true thickness can be made graphically or by calculation, if the dip of the pay formation in relation to the axis or of the bore hole can be derived from a core, from a continous dipmeter, or from some tectonic construction. Even the values for true thickness may be too optimistic, if the pay formation contains some hard, shaly or other impervious streaks and layers, which neither contain any oil or gas nor show any decent permeability. These impervious layers have to be subtracted from the true thickness of the pay, to get the value for the effective thickness, which is the only important value for the calculation of the quantities of oil or gas which are contained within the reservoir.

The porosity comprises that part of the pay formation which is open space in principle, and which can contain water, oil, or gas. The total porosity of a pay formation is called bulk porosity, and it comprises the total amount of pore space within the rock, irrespective of permeability or the situation of the pores. Some part of these pores may be closed off from the others. They can contain oil, water or gas, but by some diagenetic process they have been cut off from the pores in their neighbourhood, and their contents cannot be produced. The percentage of such pores is called dead porosity. Younger sands and clean sandstones without lime or dolomite usually have no or only very little dead porosity, and the values for their buld porosity and true porosity are identical or very similar. Limestones, dolomites and limy sandstones, however, and similar types of pay formation can have an important percentage of dead porosity, which has to be subtracted from their buld porosity for all calculations concerning the exploitation of such reservoirs.

It is important to know that most bore hole logs show the bulk porosity and not the true porosity, while most laboratory measurements show the true porosity and not the bulk porosity.

A good sand or sandstone can have a porosity of 25 or even more percent, and vugular limestones in ancient coral reefs can contain even much more pore space. Minimum pore space for gas- and oil-production is generally connected with very small permeability, but even pay formations with only ten or eleven percent of porosity can produce considerable amounts of gas, condensate or even oil.

The permeability of a pay formation is its capability to allow fluids or gases to flow in its pore space. For that purpose, the pores or at least part of them have to be connected with one another, or there must be some fissures or cracks which serve as flow channels. The bigger these channels are, the easier can a fluid or gas move, and

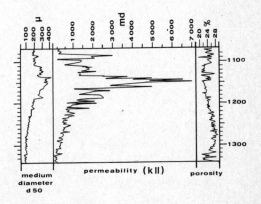

Fig. 93 The medium diameter of the sand grains, the permeability, and the porosity of a sandy pay. The grain size affects the porosity only very little, but correlates very well to the permeability (after Füchtbauer).

the better is the permeability of the pay formation. The measuring unit for permeability is Darcy or better millidarcy, after a French engineer called d 'Arcy, who published the first experiments on the permeability of sand filters. A sand or sandstone with a permeability of one Darcy or more is a very good sand, there are some better ones with even five or six Darcys, but most formations show only some hundred millidarcy. The lower limit for oil production lies towards two hundred millidarcy, bur for the production of gas even only some few milldarcy can be sufficient. These differences show that the measuring unit is independent of the medium that flows through the formation; the unit comprises the so called absolute permeability, and it is tested in the laboratory with air or an inert liquid, streaming through a small core drilled from the pay formation, or with a mercury pump.

The bore hole logs cannot be used for permeability determinations in principle, but, in general, sands and sandstones with good porosity will show a good permeability, too. Limestones and similar types of rock may be very different, and the permeability of fissured rock without real porosity can only determined by open hole or cased hole testing.

The quantitiy of fluid or gas flowing through a certain volume of rock depends on the permeability of that pay formation, on the length of that part of the pay, on the cross section or area of the part of pay in consideration, on the differential pressure which forces the oil or gas to flow, and on the vicosity of that type of oil or gas at the tempera-ture of the reservoir. The simplest formula for the flow of non-com-pressible fluids in a porous pay formation is therefore:

$$Q = K \cdot \frac{A \cdot \triangle p}{L \cdot \mu}$$

Where Q is the qantity produced within a certain time,
K is the permeability,
L is the length of that part of the pay,
A its area,
$\triangle$ p is the difference in pressure between the beginning and the end of the distance in consideration, and
μ is the viscosity under reservoir conditions.

This formula is valid only for non-compressible fluids, and only with some restrictions can it be used for oil which in general is slightly compressible. The formula is valid, too, for linear flow only, not for the radial conditions of flow around a well and of course not for the flow of gas or condensate.

Such calculations are necessary to understand the production rates of a well or of a field. They are the task of production engineers, and cannot be made without exact data from bore hole logging, from laboratory tests, and without long experience.

8.7   The Origin of Porosity and Permeability

Most porosity and permeability of pay formations is of the primary type, but there are some types of pay formation with only secondary porosity and permeability, and generally these properties have been altered and modified by diagenesis and sometimes by tectonics or even by metamorphosis.

The easiest way to understand the principles is to consider at first only a few main types, such as a sandstone, a coral limestone, and some metamorphic rock with secondary porosity.

Sands in general consist of more or less well rounded sand grains of more or less different size, which are packed together more or less closely. If the grains are well rounded, they can be washed together more closely; therefore it is important to note the grade of roundness with the description of such a pay formation, Well-rounded sand has less porosity, but about the same permeability as sands with poorly rounded, angular grains under equal sedimentary conditions.

Most important is the grain size, and the association of grain sizes. If only grains of the same or nearly the same size are present, the porosity will be high. The permeability depends on the size of the single flow channels between the sand grains. If the grains are big, the pores and with them the flow channels between them are big too, and the permeability of such a sand is high.

If the grains are small, the permeability will be correspondingly small, too. The porosity of such a sand with big grains and one with small grains can be equal, however. Porosity is affected more by the association, or the sorting of the grain sizes within the formation.

If such a sand contains as well big grains as small ones, the small ones will fill the pore space between the big grains, and the porosity will be relatively small. Well sorted sands, therefore, have relatively high porosities, and poorly sorted sands lower porosities and permeabilities.

The packing of the sand grains is another factor which affects the porosity as well as the permeability of the formation. The primary packing of sandgrains, caused by a wave on the beach, will soon be altered by physical and chemical factors, and turn the loose sand into a solid sandstone, and sometimes even into a hard and impervious quartzite. During this process of diagenesis, the porosity and the permeability is certain to became smaller, because the grains are shifted together more closely, and because minerals like calcite or iron silicates are precipitated from the pore fluids and cover the walls of the pores. In general, the porosity and the permeability decrease from the first moment of sedimentation up to the last state of meta-morphosis, but of course this procedure depends on many factors, like heat, pressure, chemistry of the ground water, and similar factors which for their part depend on the geological histroy of that pay formation. Arenites with grains consisting of calcite or dolomite instead of silica show similar trends of diagenesis and of porosity and permeability relationship. Similar trends can be found in oolithic limestones, which are built up by small balls with concentric texture inside, and which are produced by algae in flat tropical waters.

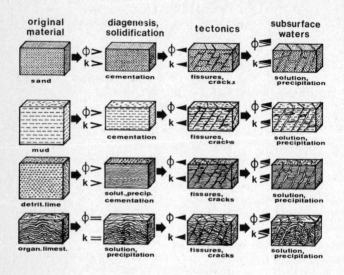

Fig. 94 The origin of porosity and permeability (k) with some different types of rock. The original porosities are modified by diagenesis, solidification, tectonics and subsurface waters in very different and specific ways.

Limestones which formerly represented a coral reef, are completely different in structure and texture, and they show very different trends of porosity and permeability relationship. The skeletons of corals contain a lot of free space inside when the soft parts of the animals are decayed. The free space between the stems and heads of the corals can be filled by debris of corals or by a more or less porous framework of other skeletons. The entire reef contains a lot of primary crevices and caverns, which provide additional porosity and permeability. Of course, the shape of such pores and flow channels looks completely different from that of a sandstone, and their geological history is very different too.

Such corals are built up mostly from calcite, with some percentage of dolomite. These carbonatic types of rock are easily soluble in sea water – easily in relation to sand grains consisting of silica – and part of that material will get into solution, and these solutions can be precipitated in other parts of that rock, decreasing the free space there. Parts of such reef limestones may become completely tight, others may show an additional secondary porosity by leaching of the sea water or ground water. The ASMARI-limestone in the Middle East, from which most of the Middle East crude originates, is such a reef limestone with very high porosity and extremely high permeability. So it is possible that wells will produce from that limestone several thousand cubic metres of crude per day, while even good wells in sandstones and sandy formations produce only one hundred tons per day or often much less.

Metamorphic rocks in general are tight and impervious. If they are crushed, however, by tectonic movements, or if parts of their minerals are leached or weathered, they may become porous and permeable. This secondary porosity is very different from the primary porosity of sandstones. Sometimes it is even difficult to measure that porosity, and to guess how much permeability there could be. Anyhow such types of rock can contain oil or gas, and some of such fields are really good producers, even if they are not of a common type.

9 A Short Course on Bore Hole Logging

Bore hole logging is one of the most common and most important methods for stratigraphic comparison, for tectonic analysis, and for pay evaluation. Modern petroleum geology and production engineering are unthinkable without bore hole logging. The principles of this method were discovered in 1921. Modern logging technique has been developed after the second World War, and even nowadays new logging methods are developed nearly every year. Every petrol company usually has at least one logging specialist. A well-site geologist should know the most important and most common types of logs, how they work, how far they are affected by the type of mud used for drilling, and other practical sides of this technique.

There are some big logging companies in the western world, like Schlumberger, Lane Wells and Atlas Gun Perforation, and a big number of smaller companies. Every company has different names for its logs, even if the systems resemble one another or are really identical.

For real understanding of the physical background and the special systems, there are several good text books. Quantitative interpretation needs some experience and special training.

9.1 Logging Car, Cable, and General Mounting

The logging car usually is a heavy truck which has to carry the cable and the cable reel, the logging cabin, the logging camera, and all the probes used for the logging program besides some other cables, rolls and similar equipment. The cable is of a very solid construction, armoured with two layers of steel wire, and with seven insulated copper cores inside. It has to withstand temperatures of more than $150°$ C in deep wells, it has to carry more than twelve tons of weight, and it has to be completely insulated even under sudden changes of pressure, in salt muds and under similar severe conditions.

The inner end of every cable core is connected to a collimator on the outside of the cable reel, and the other end fits into a cable head, which is screwed onto the probes.

The probes in general are very heavy and very solid steel or plastic tubes, barrels or bars. Some are fittet with steel springs which have to press them against the walls of the bore hole. Most of these probes contain a lot of electronic circuits, and some of them even contain radioactice sources. Many of these probes show hardly anything from the outside, they are nothing but smooth tubes or barrels. Some others, especially the probe of the electric log, show some annular electrodes of lead on a tube which seems to be made of rubber.

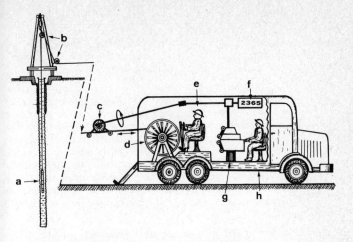

Fig. 95 The schematical procedure of borehole logging. The probe (a) is lowered into the bore hole using two or more rolls. The movement of the rope is controlled mechanically and is transferred to the camera. The driver runs the cable in and out of the hole. The operator controls the recording camera (g). The logging signals are printed on photographic film.

9.2 The Technical Procedure of Logging

One of the first difficulties is to get the logging car to the rig in time, not too early and not too late. Most logging companies have to be paid if the truck has to wait too long and on the other hand it would not be good to let the well stand still without circulation while waiting for the truck because that may cause damage to the drilling mud and to the walls of the hole. If the well is far from the next solid road one should consider that any damage happening to the truck by bad road conditions has to be paid for by the company owning the well. Every damage caused by bad well conditions to the probes, to the cable or to the truck will be charged on the company too. If a probe gets blocked in the well by a dogleg or a key hole, and if the cable has to be cut off during a fishing job, logging may become extremely expensive without comparable logging results.

During logging operations the truck should stand with its rear to the „door" of the rig, and not too far away, if possible. The cable usually goes to a big roll in the „door" of the rig, from there to another roll hanging on the hook of the travelling block, and from there is disappears in the rotary table and enters the well.

Some logging systems need electric earthing. In general it will be sufficient to use a mud tank or even the steel construction of the rig for that purpose. In regions with high resistivity layers near the surface it may become difficult to get the circuit free of electric „noise". If necessary the electricity of the entire rig has to be switched off, or the logging program has to be postponed until the next tramway or electric train has finished its activity for that night.

With nearly all logging systems some DC or AC current has to be supplied to the probe by the cable, and the logging signals come back by some other cores of the cable, enter the collector at the end of the cable, and are fed into the chest for that special logging system. The output of that chest is fed into the camera, where it moves the mirror of a galvanometer. A trace of light, coming from a stable source, is deflected by the mirror and creates a black trace on the photographic film which serves to record the log. The horizontal lines on the film which later show metres, five metres, twenty five metres, fifty metres and so on are produced by a screen,which is opened for a moment to let some light pass.

In general logging speed is about ten metres a minute, and the entire logging program for a deep well therefore will take a couple of hours, even if everything runs well. The program itself will be composed by a specialist of the company, who has the necessary knowledge and experience. The logs shall enable the well site geologist to clear the stratigraphic boundaries within the well, to locate faults and disconformities if there are any, and they are made to log the petrophysical properties of the pay, its thickness, its porosity, and its pore contents. Most logging systems are affected by the type of mud, and for every type of mud a special logging program has to be run. Thus conventional resistivity logs cannot be run i a concentrated salt mud, because that would mean an electrical short circuit within the bore hole. Special resistivity methods, with focussed patterns of ray propagation have been developed for such purposes, and high frequency logging systems have been developed for non-conducting types of mud on the other hand.

9.3 The Logging Truck

The logging truck itself contains a small cabin for the logging crew, with the stand for the driver who runs the cable in and out of the hole, the logging engineer, the logging camera and the different chests with amplifiers, transformers and other special things of the logging circuits. In general there is one chest for every logging system, and from every single system at least one trace is recorded on a photographic film in the camera.

The camera is a most complicated and sophisticated instrument, where all logging results are recorded against depth of the well together with the different scales for every type of log, with the depth

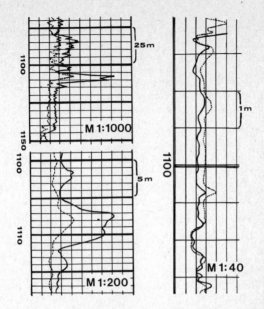

Fig. 96 The most common scales used for bore hole logging. Two different scales may be recorded simultaneously.

in metres or in feet, and with continuous or interrupted traces for better indentification. Up to four or five traces may be recorded at the same time, and if a trace is touching the border of its space on the log, a second or even third galvanometer with another scale and another amplifier will take the signal over and print its trace on the film. The rolls of photographic film on which the log is recorded, are contained in two chests on the rear of the camera. Every movement of the cable into or out of the hole in transferred to the camera by mechanical or electrical means, and during logging the film is rolled along in front of the window of the camera at a speed that corresponds to the speed of the probe within the well.

Usually two films are run at the same time, one at a scale of 1:1000 and one at 1:200, or for other purposes with 1:200 and 1:40 at the same time. When one log has been run, part of the logging cabin is transferred into a kind of darkroom, and the logging film is treated with developer, fixing solution and rinsed with water like every other photographic film. After drying heliographic prints can be made, and a head is attached to the log containing all important facts about the log and the well.

9.4 Drilling-Mud and Logging

During logging, before or after, the logging engineer will ask for a bucket full of drilling-mud, in order to measure the electric resistivity of the mud and of the mud filtrate. It would be very bad practice to take that sample of mud from one of the tanks while the mud circulation has been stopped for some hours or more. One of the most important tasks of the well site geologist or his samplers is to provide a bucket full of mud shortly before the mud circulation is stopped, and to take that sample directly from the mud return channel above or below the shale shakers. That sample will show true mud resistivities, and part of it should be run through the filter press (baroid press) in order to get some mud filtrate. Every decent logging crew will have a micro-resistivity probe for mud filtrate, where one or

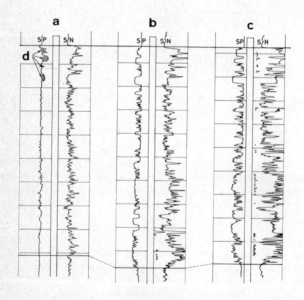

Fig. 97 The influence of the salinity of the drilling mud on the self potential log. Three wells of a gasfield, each about 1 km apart, have been drilled mit mud of different salinity and different composition. The well on the left (a) shows nearly no self potential, and reverse potentials in its uppermost part. That means that the electrical activity of the drilling mud and of the formation water are nearly identical. The well in the middle shows quite good a diagram. Its formation water has more activity than the drilling mud. The well on the right has been drilled with waterbase oil mud and shows the best diagram. The short normal on the other side of each diagram shows similar trends.

two cubic centimetres will be sufficient to get an exact date for mud filtrate resistivity. Logging engineers usually have to derive the resistivity of mud filtrate from mud resistivity by means of a chart, if they do not have any filtrate available. These charts, however, cannot be correct for all types of mud, and better results will come from real mud filtrate.

The resistivity of the mud filtrate is important for pay evaluation, because some mud will always be filtrated into every porous section of the well, and so some annular spare of the pay around the bore hole will be filled more or less completely with mud filtrate (– the so-called flushed zone –) and some part will be mixed with mud filtrate, the so-called zone of infiltration. In these zones, the resistivity of the pore contents and with it the resistivities of the pay are falsified, and do not show the „true" resistivity of a pay where the pore space is filled with oil and some connate water, or with gas, or originally with water.

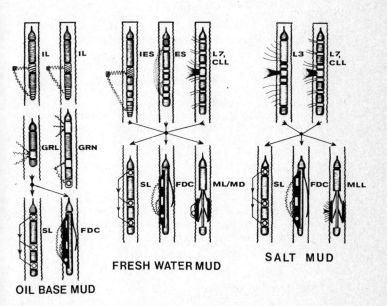

Fig. 98 Bore hole logging programs for different types of mud. For non-conductive oil base muds the induction log has to be used for logging the pore contents. Acoustical and nuclear methods may be used for logging the porosity and the shale contents of the formations and for stratigraphical comparison. Nearly all logs may be used for fresh water mud of medium conductivity. Salt muds with extremely high conductivity need special devices for resistivity logging (Abbreviations see p. 155).

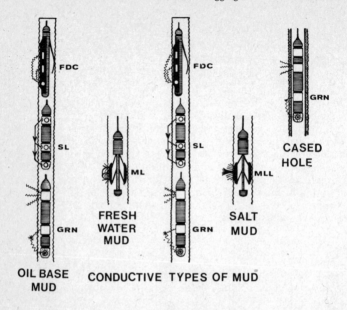

Fig. 99 How to log the porosity of a pay formation. For non-conductive types of mud only nuclear and acoustical methods may be used. In fresh water mud the microlog may be used moreover. The microlaterolog has been developed especially for salt mud. In cased holes, only the gammaray neutron will give good results (Abbreviations see p. 155).

9.5 The Logging Systems

In course of time, a long series of logging systems and methods have been developed, and it would lead too far to mention all the names here. In principle there are electric methods, nuclear methods, acoustic methods, and others.

9.6 Electric Logging Systems

Most electrical methods log the *resistivity of the rock* around the bore hole. The specific resistivity shows what type of rock it may be in general, how many conducting parts and how many nonconducting parts like lime, silica, and so on it contains. Shale in general has a very low specific resistivity, while lime, limy sandstones and quartzite have a relatively high resistivity. Porous sandstones, however, containing high salinity water in their pore space may show very low resistivities too. Oil and gas within the pore space of a sand or sand-stone will increase its resistivity because they are non-conductive materials. It will show lower conductivity or alternatively higher resistivity on the diagram.

The so-called convential resistivity logs show a strict interdependence of their depth of penetration and their resolution of petrographic details. If a deep penetration is needed to log the pay out of the zone of mud filtration, the resolution is very bad, and vice versa.

Better results are shown by the so-called focussed methods where the logging current is compressed to a trumpet or to a disk by a physical trick, and is forced to penetrate relatively deep into the rock at a good rate of resolution.

For all types of drilling fluid with bad conductivity, such as oil base muds, and even for holes drilled with gas or compressed air, special high frequency methods have been developed, called Induction Log, High Frequency Log, or similar. This type of log shows no outer electrodes. It contains an emitter coil, which sends the logging current into the formation and one or several receiver coils, which measure the response of the formation and transfer it to the surface by means of the cable.

This type of logging tool may be used in combination with conventional resistivity methods, or even with focussed logging arrays.

There are two methods of logging the *porosity of rock* by resistivity methods, too, the so-called microlog and the microlaterolog, with similar names when sold by other companies. These logs measure the resistivity of the annulus of rock which surrounds the bore hole, where all or nearly all original pore contents have been replaced by mud filtrate. The logging signal then contains the resistivity of the non-conducting solid parts of rock, and the conductance of the mud-filtrate in it. The resistivity of the mud filtrate is known by the sample of mud checked before logging and the porosity of that part of rock may be determined by means of a simple calculation or, better, by means of a chart developed for that purpose.

Another electric tool logs *natural potentials* within the bore hole, which develop around porous zones. The theory of these spontaneous potentials is very complicated and doubtful, but the diagram is very easy to read when the bore hole has been drilled with a mud of poor salinity, and the pore contents are of higher salinity or vice versa.

Deflections of the diagram from a so-called „shale line" indicate porous zones. These deflections show to the left side, if the salinity of the pore contents is higher than that of the drilling fluid, or alternatively they may deflect to the right when soft water-bearing formations are drilled through with mud of higher salinity.

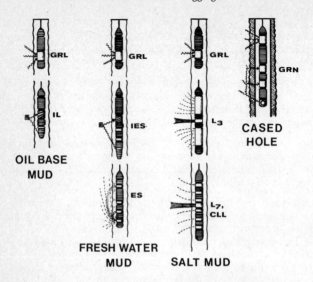

Fig. 100 How to log the pore contents of the pay formation. For non-conductive muds the induction log and the gamma ray log may be used. In wells drilled with fresh water muds the gammaray will show the shale contents and the electriclog and the induction-electriclog will show the water saturation. In salt muds only focussed devices will show the water saturation. Cased holes may be logged only by gammaray-neutron.

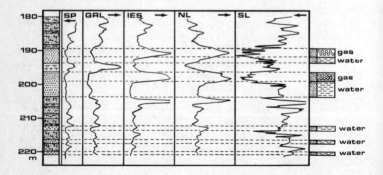

Fig. 101 A set of diagrams from a gas well. The soniclog shows the porous zones very clearly. The neutron log shows the zones with gas and the gammaray log shows the difference between sand and shale. The induction-electriclog corresponds to the neutronlog, showing the gas bearing parts of the pay. The selfpotential is not very clear.

9.7 Nuclear Logging Systems

Nuclear methods have the great advantage of being nearly independent of the type of mud. They use very different methods and principles, and nearly every year a new type appears on the market. The oldest and most usual type is the Gamma Ray Log, or *Natural Gamma Log*. The probe usually contains a scintillometer, and the diagram shows the gamma-activity of the rock, recorded against the depth. This natural gamma activity mostly dates from a radioactive isotope of potassium, called potassium – 40, which is contained at a very low percentage in all types of shale, mudstone, and in marls and limestones according to their shale contents. The only sources of error may be monazite sand containing thorium minerals, and black marine or brackish shales containing salts of uranium. These two possible sources of error, however, are so rare, that they need not be taken into account. In general, the radioactivity logged by the gamma ray log shows the shale contents of the rock, and that is a very simple but useful log.

The scintillation crystals used for such logs are sensitive to heat. In hot bore holes, that means at temperatures exceeding about $120° C$, they usually „get crazy". They develop false readings, and the whole diagram will shift over to the right and leave the scale.

In such bore holes, the old Geiger-Mueller-type of gamma probe has to be used, which is not affected by heat, buth which is much less sensitive, and therefore has to be run at a speed of less than three metres/minute.

The *Density Log* has been developed to determine the porosity of rock, the compaction of shales and similar details by nuclear methods. The probe has to be pressed against the bore hole wall by an elastic spring. It contains a source for gamma radiation and a scintillation detector. The gamma radiation is directed towards the bore hole wall by a narrow collimator, and the gamma rays which are scattered back from the formation and which enter the collimator of the scintillation detector ar counted. The counting rate corresponds to the density of the formation.

The *Neutron/Neutron Log* logs the contents of the formation in light nuclei, especially the hydrogen contents. The probe contains a source which emits fast neutrons, and these fast neutrons are slowed down by collisions with light nuclei. Gradually they be come thermal neutrons, and these thermal neutrons may be counted by a special counter, containing a scintillation crystal made from lithium fluoride. The counting rate then corresponds to the contents in water, and to the porosity. Gas-bearing formations need special considerations.

The *Neutron/Gamma Log* contains the same source which emits fast neutrons. The counter, however, is sensitive to so-called gamma rays of capture. These highly energetic gamma rays develop when neutrons are slowed down by continuous collisions with light nuclei till they become subthermal neutrons. Then they may be „caught in" by the nucleus of a hydrogen atom, and a gamma ray of capture will get free. The distance between source and counter is bigger than with the neutron/neutron log, and the penetration into the formation is somewhat greater. In general the counting rate corresponds to the porosity, as with the neutron/neutron log.

During recent years some new types of nuclear logs have been developed with Californium-252 as a source. This element emits fast neutrons too, but at a greater rate of activity than the usual Am/Be sources. Elements which are kept long enough under the irradiation of these fast neutrons become radioactive themselves. The energy level of their radiation is different, and a kind of chemical analysis can be made using multi-channel counters. So e.g. the contents of silicium and aluminium in the formation may be counted by a two channel analyser, and that ratio will correspond to the content of sand and shale in the rock.

9.8　Acoustic Logging Systems

Acoustic logs either measure the velocity of sound within the formation, or the attenuation of sound. The velocity of sound is specific to all kinds of rock, and so the type of rock may be detected by such velocity diagrams. The velocity of sound within a porous sandstone or limestone, however, will depend on the grade of porosity, and therefore these velocity logs are used mainly for porosity determinations. The probe contains an emitter, which produces pulses of ultrasonic sound, and two or more receivers. The time of travel between two such receivers is measured and plotted against the depth. Most porosity tools use receivers at a distance of one foot, for seismic purposes even tree-foot tools are used. Attenuation logs are used for two completely different purposes, for the detection of fissures, crevices and cracks in solid rock, and for the control of cementation. Every change from solid rock material to water, oil or gas within a fissure will attenuate the pulse of sound sent across it, and the more fissures there are, the lower will the amplitude of the pulses become during their travel from the emitter to the receiver. Such fissured formations sometimes contain oil, and the other porosity logs will not give any good information on their contents in free space.

The cementation of the casing should have a good bond with the casing itself as well as with the formation. The so-called cement bond log measures the attenuation of sound on its path from the bore hole across the casing, the sheet of cement and the formation and back again to the reciever in the bore hole. If there is a good bond everywhere, the attenuation will be small. If the attenuation is higher than normal, the bond will be imperfect.

9.9 Some other Types of Logs

The deriviation logs contain a compass and a tiltmeter. Both sensors produce a log, and the deriviation of the borehole at every depth can be calculated from that log. Of course it is not possible to measure the direction or the tilt while the probe is in movement, it has to be stopped every twenty metres or so. Older types of derivation log use a kind of photocamera which takes a picture of the compass and of the pendulum at every point. Some logging companies will not use a real pendulum, but a small steel ball running free on a convexe glass plate supplied with concentric rings.

The calibre log may be run together with sonic, electric, or nuclear devices. Generally it consists only of one or several springs which slide along the walls of the bore hole, and of a simple electric apparaturs which measures the angle between the body of the probe and the spring or springs. These calibre logs are very useful, not only for technical control, but mainly for the detection of soft formations which have been caved out during drilling, of solid formations which retain their calibre, and of porous formations which show a narrower calibre because their surface is covered by mud cake.

9.10 Stratigraphic Correlation of Bore Hole Logs

One of the most important tasks for the well site geoligist is to provide a proper and exact stratigraphic correlation of the newly-drilled well with the other wells in that field, or with wells in the neighbourhood if the well is a wildcat. For that correlation only diagrams on a scale of 1:1000 or similar should be used, and always the same types of log if possible. These diagrams of the nearest wells are pinned against a wall, a board, or they are put on a glass plate. In most cases it is best to use not the original logs, but the so-called completion logs, which contains all important stratigraphic limits and other important facts. The new log is pinned between the other logs, and the first work is to compare the diagrams from some distance. Then they are shifted until the main units resemble one another, and finally every peak on the new diagram is compared with every peak on the other logs. Many of these peaks have a characteristic shape, or some groups of peaks return in these logs always in the same stratigraphic sequence. In general it needs some time and experience in every region to learn this technique. In marine basins, correlation in principle is much easier than in aestuarine, brackish, limnic or terrestrial facies. Some peaks, representing some special marine facies, may be traced from log to log over hundreds of kilometres, while some big aestuarine conglomerates may disappear from one well to the next, and if they reappear, they often are younger or older, or contain other material. All that is very logical and easily understandable, if the conditions of sedimentation in these environments are taken into consideration.

One of the principal laws for such correlation is:

The coarser a sediment is, the less it can be used for strati-graphic correlation.

It should be clear that the height of peaks can be affected by the influence of the drilling mud, by the size of the hole, or by other factors depending on the type of log. Electric logs are very sensitive to the resistivity of the drilling mud, especially the Electriclog or other non-focussed arrays. If such diagrams are used, one should not consider the height of the peaks in comparison to the scale, but only their height in comparison to adjacent peaks, or the trend of increasing or decreasing resistivity.

Care should be taken, however, not to use the so called Lateral or Inverse curve among the resistivity logs. This curve is reliable only if a pay or a layer has a certain thickness, and if there are no layers of high resistivity above that peak. Anyhow it is better not to use that diagram, because peaks on that curve may appear or disappear only because the thickness of that layer has changed a bit, or because the resistivity of a layer some metres above has changed.

9.11   How to Trace a Fault or a Disconformity by Bore Hole Logs

In principle, faults will not show any direct trace or signal on a log, they have to be traced by the gap or by the duplication of parts of the formations in the bore hole, and by parts within the diagram which are either lacking or else duplicated. It needs some exact work and some experience to detect a fault within a log, and to determine how many meters of formation are lacking. The best way is to put the log of the new well and the log of the wells in the neighbourhood on a table with a transparent plate and some neon tubes below. Of course only curves with best resolution should be used, like e.g. the Short Normal, a focussed resistivity log, a one-foot acoustic log, or similar devices.

First the approximate situation of the fault may be traced by a coarse comparison of several peaks of the diagrams, and their distance from one another. That region is market by a big circle or the like, and then two diagrams are put one on the other so that the two curves nearly cover one another. The new diagram should be lying on top, and at first the peaks are compared from above to below, until they do not correspond any longer. The last „sure" correlation is marked with a pencil, and then the diagrams are shifted, so that they correlate in their lower part. The peaks are compared again from below to above, and the end of sure correlation is marked again on the diagram. If both marks are identical, they show the position of the fault. If they are some metres apart, the fault must be somewhere between them.

The thickness of that part of the formation which is lacking or which appears twice, once above and once below the true fault may be determined in the same way taking the last „sure" peaks for comparison. Of course such a gap within a sequence does not necessarily mean a fault; it may be caused by a disconformity, too. The effect of a disconformity is the same, but in general such disconformities are known, and will not be mixed up or confused with faults. In unknown regions, however, they may cause some doubts and difficulties. The best way to clear up such problems is to make a fence diagram or a stereoscopic diagram of that oil field or that region. Disconformities usually occur always in the same or nearly the same stratigraphic position, and they can be distinguished easily from faults if their angle of dip is compared with the adjacent beds.

There is one type of log which may show even such tectonic features as faults and disconformities, the so called dipmeter or continuous dipmeter. This probe contains a compass, an electric pendulum, and three or four electrodes which measure the resistivity in one plane perpendicular to the bore hole. The four electrodes enable the operator to calculate the apparent dip of the formation around the bore hole, and the compass and pendulum give the necessary orientation and the correction for the tilt of the bore hole axis. Disconformities show up by the different dip of the beds above and below them, and faults may be traced, if the blocks above and below the fault are different in dip.

9.12 How to Evaluate a Pay from Bore Hole Logs

In general the evaluation of a pay by bore hole logging needs a lot of knowledge and experience. It should be done therefore by trained specialists only, who know about the difficulties and the possible errors. This evaluation has three main tasks, the determination of the effective thickness, the determination of the grade of porosity, and the determination of the pore contents.

The effective thickness of the pay comprises only its porous parts. They may be derived from the caliper log, the microlog, the density log, the sonic log or similar methods. In all formations which contain no lime or limy sandstones, a beginner will do best to use a

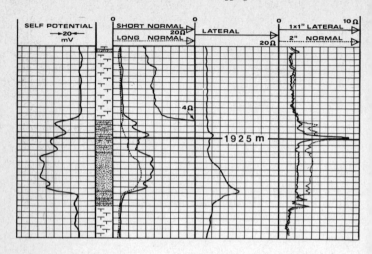

Fig. 102 A combination of diagrams from an oil well. The self potential shows the bulk thickness of the pay. The short normal and the long normal point out that the resistivity of the pay formation is much higher than that of the shale below and above the pay. The lateral curve indicates a resistivty of about 10 ohms/m. The microlog shows a hard streak in the upper part of the pay formation.

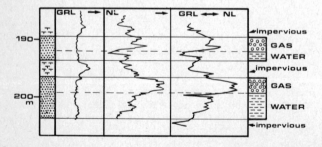

Fig. 103 A combination of logs from a well of an aquifer storage field. The first set of diagrams, a separate gammaray diagram (GRL) and a neutron log (NL) have been run only short time after the begin of storage. The combination of gammaray and neutron shows some small differences but in general the same distribution of gas and water.

microlog for that part of the evaluation. All other logs and especially all limestones and limy sandstones need some more precautions.

The porosity of these porous parts may be determined by a density log, a neutron log, a sonic log, a microlaterolog, or even by a microlog. A microlog should not be used for that work if the pay tends to lime or to limy sandstone. Other difficulties and errors may arise from shaly sands and sandstones.

The pore contents are somewhat difficult to determine, expecially in regions with soft or low grade brackish waters. For this determination the resistivity of the pay is used as a criterion. Marine formation waters have a very low resistivity, whereas pays filled with oil and gas have a relatively high resistivity. For such regions and for pay zones devoid of lime there is a very simple „rule of thumb" which can be used: If the resistivity of the pay is higher in the Long Normal than in the Short Normal curve, the pay will contain oil or gas. If the Short Normal shows higher values than the Long Normal, the pay will contain water.

In regions with soft or brackish formation water this rule of thumb will not work, and sometimes a test has to be run to be sure if the pay really contains water or hydrocarbons. Oil and gas show only very small differences in resistivity logs, and they have a different effect on nuclear and on acoustic logs.

In existing oil- and gasfields it will be possible to determine whether a newly drilled well is worth completing or not. With wildcats in little known regions, however, nobody will fill the hole up as long as there is any doubt about the contents of the pay, and before the well has been tested.

Explanation of Abbreviations:

ES	=	electric log
IL	=	induction log
IES	=	induction electric log
L 3, L7, CLL	=	focussed electric logs
ML	=	microlog
MLL	=	microlaterolog
MD	=	microdiameter
GRL	=	gamma ray log
GRN	=	gamma ray neutron log
FDC	=	formation density log
SP	=	spontaneous potential log

10   Open Hole Testing

Open hole tests have to clear if it is worth running the final casing into a borehole and completing it, or if it has to be abandoned. There are two ways of testing a well without casing, drill stem tests, and cable tests.

The *drill stem tester* is run into the hole using the usual drill pipes. It consists of a complicated system of valves and pressure sensors, and there are several different versions possible depending on the distance between the pay to be tested and the bottom of the hole.

If that space is less than about thirty metres, the tester can be set up on a tail pipe that stands on the bottom of the hole. This tail consists of normal drill pipes, and of a perforated liner. When the tester is set up, the weight of the drilling string above the tester inflates a rubber packer and the part of the bore hole below the tester is sealed off from the rest of the bore hole. Then a valve within the tester is opened by rotation of the string, and this part of the well with the pay formation is opened to the interior of the drilling string. The reservoir pressure is then no longer balanced by the weight of the mud column, and the contents of the pay formation are forced to flow into the tester and the test string. After some twenty minutes of testing the valve is closed again, the tester and the testing string are run out of the hole, and the contents of the string can be analysed. Gas, of course, will blow out and has to be flared off in order to avoid explosions. If the pay contains oil, it will always be mixed with mud filtrate. Such mud filtrate, however, and formation water may be distingushed by their different contents in salt and other chemicals.

If the distance between the pay and the bottom of the well is greater than about thirty metres, it is only possible to run a *straddle test,* or to run several straddle tests if there are several pay zones.

Such straddle testers contain two rubber packers which can seal a certain part of the well off from the rest of the bore hole. The tester is set off by means of a so-called hook wall anchor, a system of steel grips which are run into the walls of the bore hole and keep the packers above or below the pay zone to be tested. The perforated liner is between the two packers, the rest of the string is similar.

The function of such a straddle test or of a single test with a foot anchor is checked continously by a set of pressure sensors which are built in at different parts of the tester, and which record the varying pressures on a built-in chart. Modern testing strings enable the operator to vary between the closed-in state and the flow state of the test. During the closed-in state the tester is open to the pay zone, but the valve that opens the exit into the drillpipes and to the surface is still closed. The diagram then shows the formation pressure. When the flow valve has been opened, the pressure within the tester will drop, showing the lower flow pressure around the bore hole. After

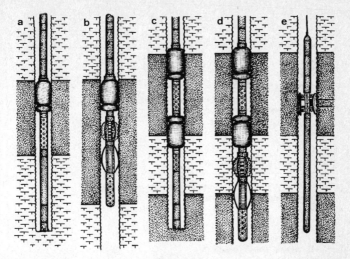

Fig. 104 Different possibilities of open hole drill stem tests. a) Single test with bottom anchor. b) Single test with hook wall anchor. c) Straddle test with bottom anchor. d) Straddle test with hook wall anchor. e) Cable tester.

some time of flowing the flow valve is closed again, and the pressure diagram will more or less slowly regain its former state. Pay formations with good permeability will fill the gap very quickly, and bad pay formations with low permeability will do so only very slowly. Such an open hole test already gives a certain picture of the production which can be expected from that well.

Such drill stem tests have the disadvantage that they need a lot of time, and that the seal between the packer and the pay formation is not always tight. Tests on the logging cable, using a so called cable tester, have the big advantage of needing only very little time for running in and out. In principle, they can be made just after the logging program, with the same logging truck, and on the same logging cable.

The tester is a very sophisticated tool which works according to push-button signals from the surface, while a pressure diagram is recorded on the usual camera, and a sample of the formation fluid is preserved in a container hanging below the tester.

When the tester is opposite the pay zone, which may be controlled by a self-potential curve, it is pressed against the bore hole wall by a packer shoe, and one or two shaped charges shoot flowing channels into the formation. These flow channels are blocked off from the bore

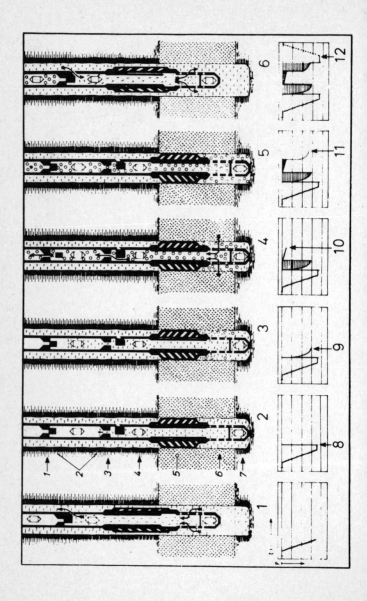

hole by a rubber seal, and after some time of shut-in pressure testing, another valve can be opened, and the contents of the pay formation flow into the container below the tester. Thereafter, the flow valve can be shut again, another shut-in pressure diagram can be recorded, then the packer shoe is released, and the entire tester is drawn to the surface. There the contents of the container can be analysed. The advantages of this system are obvious, especially in respect of the short testing time and the instand results. A certain handicap is the restricted volume of the container. In formations with good permeability it may happen that only or mostly mud filtrate flows into the container, while the original pay fluids do not have a chance of reaching the bore hole. Care should be taken therefore to use the biggest container possible.

Fig. 105 Open hole drill stem test (single test with bottom anchor). Schematic drawing of test phases and pressure diagrams of a pressure recorder outside the flow chamber. 1. Running in the tester. Mud is equalized flowing through liner and spring valve. Graph shows pressure increasing with depth. 2. Packer set off. Spring valve opens flow chamber. Graph shows drop of initial pressure. 3. Initial close-in-pressure test. Graph shows pressure build-up. 4. Flow test. Auxiliary valve above flow chamber is open. Pay fluids flow into string of drill stems. Graph shows flow pressure build-up. 5. Final close-in-pressure test. Auxiliary valve is closed. Graph again shows final close-in-pressure build-up. 6. Test finished. Spring valve is reopened to mud column, mud equalizing through packer. Graph shows mud pressure decreasing with depth. Legend, 1 – auxiliary valve, 2 – flow chamber with pressure recorder, 3 – spring valve, 4 – pressure recorder within flow channel, 5 – packer, 6 – liner, 7 – packer shoe with pressure recorder, 8 – initial mud pressure, 9 – initial close-in-pressure, 10 – flow pressure, 11 – final close-in-pressure, 12 – final mud pressure.

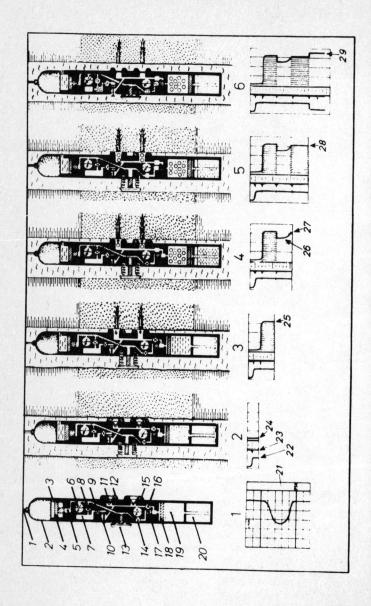

Fig. 106 Formation-Interval Cable Tester (FIT) (Schlumberger). Schematic drawing of test performance and test graphs. 1. Schematic build-up of cable tester; situation before the test. 2. Tester set off. Mud valve opened, mud column working on the piston. Piston works on hydraulic pressure line and shifts the supporting packer against the borehole wall. 3. Initial close-in-pressure test. Pay is opened by two shaped charges, close-in-pressure building up within flow line. 4. Flow test. Flow valve is open, pay producing into sample container. 5. Final close-in-pressure test. Sample valve is open; hydraulic line working on sample piston and closing sample container, flow line is closed. Final close-in-pressure building up. 6. Tester is free. Both valves for equalizing hydraulic pressure are open, supporting packer is drawn back to tester, freeing valve is opened and tester gets free from bore-hole wall. Legend, 1 – cable, 2 – cartridge with electrical set up etc., 3 – mud valve, 4 – mud piston, 5 – hydraulic line, 6 – pressure recorder for hydraulic line, 7 – pressure equalizing chamber, 8,9 – pressure equalizing valves, 10 – freeing valve, 11 – shaped charge, 12 – ring packer, 13 – supporting packer, 14 – sample valve, 15 – pressure recorder in flow line, 16 – flow valve, 17 – sample piston, 18 – floating piston with sample container, 19 – water cushion, 20 – air cushion, 21 – SP or GRL diagram for setting tester, 22 – pressure for hydraulic line, 23 – graph of electrical function. 24 – pressure within flow line, 25 – initial close-in-pressure, 26 – flow pressure, 27 – beginning of the final close-in-pressure with filled sample container, 28 – final close-in-pressure, 29 – mud pressure.

11   What a Geologist Should Know about Well Completion

A well containing oil or gas has to be made ready for production, the casing has to be run in, the producing zone has to be perforated, the wellhead has to be installed, the oil pump or bottom chokes have to be run in witch the tubings, and perhaps a production packer has to be installed. All that is merely technical work, it will be controlled by some reservoir engineer, and it will be done by the roughnecks of the rig that drilled the well or by the personnel of the oil field.

Every well site geologist, however, should at least understand what is happening, and why it is done.

The casing consists of steel tubes which are screwed together. In gas fields a special type of thread has to be used which gives a tight seal even for gas. The string of casing is run into the hole almost like a string of drillpipe, but with different power tongues and other wedges.

The lower end of the string contains the cementing shoe, made partly from concrete or light metal. As soon as the string is run in, it is cemented in order to get a tight seal between the pay formation and any other formations containing some other fluid or gas. The cement is mixed with water and pumped down after a rubber plug which separates the cement from the drilling mud in the hole. When this plug reaches the cementing shoe, a rupture disc on its surface is broken by the pumping pressure, and the cement slurry begins to fill the space between the casing and the bore hole walls. So-called centralisers have to keep the casing in the centre of the hole to provide a good, uniform seal.

After the cement slurry a second rubber plug is run into the hole and is pumped down with water or mud. This plug will come to rest on the first one, and will seal the hole hermetically. In earlier times it was usual to set the casing off on top of the pay zone, and to let the pay open, if it was solid enough, or to open it by a so-called liner, a tube with slits or holes. These methods have meanwhile been abandoned. Usually the casing is run below the pay zone, and the necessary perforations are shot into the casing using special perforators.

These perforators are run into the hole on a logging cable and shoot the perforations into the bore hole walls by means of steel bullets or shaped charges. The old type perforators are heavy steel bars containing the bullets and/or the shaped charges. They may be used in bore holes only before the tubings have been run in.

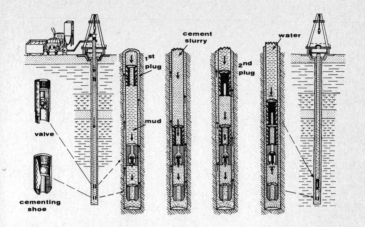

Fig. 107 The procedure of cementing a string of casing. The cement slurry is pumped in after the first plug. As soon as this plug arrives at the cementing valve, a rupture disk on its surface is broken by the pumping pressure and the cement slurry begins to fill the space between the casing and the formation. The second plug is run in after the slurry. It is pumped down with water. Cementing is finished, when both plugs have arrived above the valve.

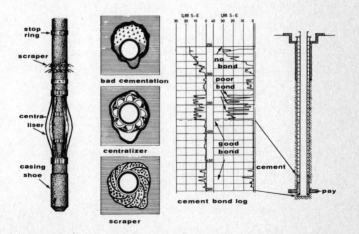

Fig. 108 How to get better cementing. Centralizers keep the casing in the center of the hole. Scrapers cleen the walls of the bore hole. The cement bound log (CBL) shows the quality of the cementation.

Most modern perforations are made by shorter or longer strings of capsules containing shaped charges. The debris of the string itself will fall down to the well bottom and remain there.

The tubings are narrow, thinwalled tubes of about two or three inches width. They can be installed with their lower end hanging free in the casing, or that end may be run into a so-called production packer, to get a tight seal between the casing and the string of tubings.

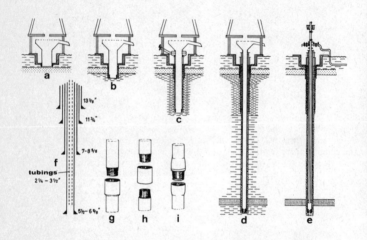

Fig. 109   Different types of pipe, casing and tubings used for a well. a) The mud funnel, and b) the stand pipe. c) The gas anchor, and d) the final casing. e) The well after completion with tubings and well head. g, h, i) Different types of casing. Extreme line casing (i) is used especially for gas wells.

If an oil well has to be pumped, the oil pump is inserted into a special seat within the tubings, from where it can be or removed or released for repair or cleaning. Such an oil pump is usually a very simple tool with two ball valves and is driven by a string of relatively tiny steel bars.

These steel bars are moved up and down by the so-called horse heads on the surface, which in their turn are driven by electric motors.

Oil wells with a good energy level usually flow without pumping. Their production is controlled by one or more chokes, which can be shut into the flow path by corresponding valves. The upper ends of the outer casing, of the inner casing, of the tubings, and these

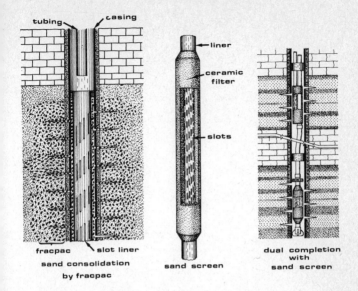

tubing   casing

liner

ceramic filter

slots

fracpac   slot liner

sand consolidation
by fracpac

sand screen

dual completion
with
sand screen

Fig. 110 The consolidation of loose sands with fracpac, and the completion of such a well with a so called sand screen.

branches with different chokes with manometres and other control instruments look like a big double or triple cross made from steel tubes. This part of the completion is called the eruption cross, or more usually the „Christmas tree".

When the Christmas tree is installed, and the new well produces its oil or gas into a pipeline or a reservoir tank, the work of the well site geologist is done, and he well drill for oil or gas somewhere else. Perhaps he has to draw some profiles which show the situation of the newly drilled well within the field or the general structure. This work, however, is already part of what is usually calles exploitation. It is not possible to draw an exact line between exploration and exploitation, but in general exploitation begins as soon as the well begins to produce. All these special tasks connected with producing oil or gas are called „production engineering" in general, and will be treated in a special pamphlet of the same series of booklets.

Literature

Abelson, P. H.: Organic geochemistry and the formation of petroleum. - Proc. 6th World Petroleum Congress, Frankfurt 1963, Section 1.

Adams, J. S. & Gasparini, P.: Gamma-ray spectrometry of rocks. - Elsevier 1970.

Alliquander, O.: Das moderne Rotarybohren. - VEB Verlag, Leipzig 1965.

Amico, M.: Petrolio e gas naturale. - Hoepli, Milano 1953.

Antoine, J. W.: Geology and hydrocarbon potential, deep Gulf of Mexico. - Bull. Amer. Assoc. Petrol. Geolog., **54**/5, 834, Tulsa 1970.

Apostolescu, V.: An attempt at zoning by ostracods in the Cretaceous of the Senegal Basin. - Rev. Inst. Franc. Petrole, **18**, 1675, Paris 1963.

Arps, A. A.: Estimation of primary oil reserves. - Trans. AIME, Petroleum Transactions, **207**, 182, 1956.

Arps, A. A.: Estimation of primary oil and gas reserves. - In: Petroleum Production Handbook. - Ed. T. C. Frick, **2**, 37, 1962.

Auboin, J.: Géosynclines. - Elsevier 1965.

Baker, E. G.: A geochemical evaluation of petroleum migration and accumulation. - In: Fundamental aspects of petroleum geochemistry. - Elsevier, Amsterdam 1967.

Beck, R. H. & Lehner, P.: Oceans, new frontiers in exploration. - Bull. Amer. Assoc. Petrol. Geol., **58**/3, 376, Tulsa 1974.

Beckmann, H.: Einfluß der Spülung auf Bohrlochmessungen. - Erdöl-Zeitschrift **79**, Wien 1963.

Beckmann, H.: Geologische Bedeutung der gasmeßtechnischen Spülungsüberwachung von Erdölbohrungen. - Bergbau-Wissensch. **10**/11, Goslar 1963.

Beckmann, H.: Modern methods of pay evaluation in oil and gas fields. - Rudarsko-Metallurski Zbornik **4**, Lubljana 1964.

Beebe, B. W. & Curtis, B. F.: Natural gases of North America. - Amer. Assoc. Petrol. Geol., Tulsa 1968.

Bitterlich, W. & Wöbking, H.: Geoelektronik. - Springer Verlag, Wien 1972.

BP Benzin und Petroleum AG: Das Buch vom Erdöl. - Reuter und Klöckner Verlag, Hamburg 1967.

Bray, E. B. & Evans, E. D.: Hydrocarbons in non-reservoir rock source beds. - Bull. Amer. Assoc. Petrol. Geolog., **49**/3, 246, Tulsa 1965.

Breitenbach, E. A.: Computer evaluation of logs. - Journ. Petrol. Technol. **18**/4, 493, 1966.

Brown, A. A.: New methods of characterizing reservoir rocks by well-logging. - PD **3**/2, 7 World Petrol. Congress, Mexico 1967.

Burk, C. F.: Computer-based geological data systems: an emerging basis for international communication. - RP **5**, 8th World Petrol. Congress, Moscow 1971.

Cagniard, L.: Basic theory of the magneto-telluric method of geophysical prospecting. - Geophysics **18**/3, 605, 1953.

Caioli, L.: Tecnologia della perforazione, ricerca e coltivazione dei minerali fluidi. - Hoepli, Milano 1951.

Chapman, R. E.: Petroleum geology. - Elsevier, Amsterdam 1973.

Carrigy, M. A.: The physical and chemical nature of a typical tar sand. - 7th World Petrol. Congr., Mexico, PD **3**/2, 31, 1967.

Chandler, G.: Energy; the changed and changing scene. - Petrol. Rev. **165**, 1973.

Chapman, R. E.: Petroleum Geology. - Elsevier, Amsterdam 1973.

Chapman, R. E.: Petroleum and Geology: a synthesis. - APEAJ, **12**/1, 36, 1972.

Choquette, P. W. & Pray, L. C.: Geological nomenclature and classification of porosity in sedimentary carbonates. - Bull. Amer. Assoc. Petrol. Geolog., **54**/2, 207, Tulsa 1970.

Cooper, C. C.: Experience with the Permian basin well-data system. - World Oil, **82**, April 1967.

Cordell, R. J.: Colloidal soap as proposed primary migration mecanism. - Bull. Amer. Assoc. Petrol. Geol., **57**/9, 1618, Tulsa 1973.

Cordell, R. J.: Future of geology in petroleum exploration. - Bull. Amer. Assoc. Petrol. Geol., **52**/2, 475, Tulsa 1968.

Cox, B. B.: Transformation of organic material into petroleum under geological conditions. - Bull. Amer. Assoc. Petrol. Geolog., **30**/5, 645, 1946.

Cross, A. T.: Palynology in oil exploration. - Soc. Econ. Paleontolog. Mineralog. Spec. Publ. **11**, 29, 1964.

Dakhnow, V. N.: Geophysical well logging. - Colorado School of Mines **57**/2, Golden, Colorado 1962.

Dallmus, K.: Mechanics of basin evolution. - In: "Habitat of oil." - Amer. Assoc. Petrol. Geolog., Tulsa 1958.

Desbrandes, R.: Théorie et interprétation des diagraphies. - Editions Technip, Paris 1968.

Dobrin, M. B.: Introduction to geophysical prospecting. - McGraw-Hill, New York 1952.

Domzalski, W.: Airborne techniques in petroleum exploration. - Journ. Inst. Petrol. **48**/459, 55, 1962.

Dott, R. H. & Reynolds, M. J.: Sourcebook for petroleum geology. - Mem. **5**., Amer. Assoc. Petrol. Geolog., Tulsa 1969.

Duntun, M. L. & Hunt, J. M.: Distribution of low molecular weight hydrocarbons in recent and ancient sediments. - Bull. Amer. Assoc. Petrol. Geolog., **46**/12, 2246, Tulsa 1962.

Engelhardt, W. v.: Der Porenraum der Sedimente. - Springer-Verlag, Göttingen 1960.

Erdman, J. G.: Petroleum – Its origin in the earth. - In: Fluids in subsurface environments. - Amer. Assoc. Petrol. Geolog., Tulsa 1965.

Faniev, R. D.: Abbau von Erdöl- und Erdgaslagerstätten. - VEB Deutscher Verlag, Leipzig 1963.

Fisk, H. N.: Bar-finger sands of Mississippi delta. - In: Geometry of sandstone bodies. - Amer. Assoc. Petrol. Geolog., Tulsa 1961.

Flores, G.: Introduzione alla geologia del petrolio. - Flaccovia Edit., Palermo.

Forbes, R. J.: Studies in ancient technology. - E. J. Brill, Leiden 1955.

Forbes, R. J.: Studies in early petroleum history. - E. J. Brill, Leiden 1958.

Forgotson, J. M. & Stark, P. H.: Well data files and the computer; a case history from northern Rocky Mountains. - Bull. Amer. Assoc. Petrol. Geolog., **56**/6, 1114, Tulsa 1972.

Frankel, P. H. & Newton, W. L.: Comparative evaluation of crude oils. - Journ. Inst. Petrol. **56**/547, 1–8, 1970.

Gavat, J.: Geologia petrolului si a gazelor naturale. - Edit. Didactica si Pedagogica, Bucuresti 1964.

Gill, D.: Application of a statistical zonation method to reservoir evaluation and digitized log analysis. - Bull. Amer. Assoc. Petrol. Geolog., **54**/5, 719, Tulsa 1970.

Glaessner, M.: Micropaleontology. - Melbourne Univ. Press, 1945.

Gotautas, V. A.: Quantitative analysis of prospect to determine wether it is drillable. - Bull. Amer. Assoc. Petrol. Geolog:, **47**/13, 1794, Tulsa 1968.

Graaf, R. M. van der: Recent advances in formation evaluation. - RP **5**, 7th World Petrol. Congr., Mexico 1967.

Graaf, R. M. van der: Application of computers in formation evaluation. - Journ. Inst. Petrol., **54**/540. 380, 1968.

Gregory, A. F.: Analysis of radiometric sources in aeroradiometric surveys over oilfields. - Bull. Amer. Assoc. Petrol. Geolog., **40**/10, 2457, Tulsa 1956.

Griffiths, D. H. & King, R. F.: Applied geophysics. - Pergamon Press, London 1965.

Gussow, W. C.: Salt diapirism; importance of temperature and energy source of emplacement. - In: Diapirism and Diapirs. - Amer. Assoc. Petrol. Geolog., Tulsa 1968.

Gutjahr, C. C. M.: Carbonisation measurements of pollen grains and spores and their application. - Leids. Geol. Medl, **38**/1, 1966.

Halbouty, M. T. & Meyerhoff, A. A. & King, et al.: World's giant oil and gasfields. - In: Geology of giant petroleum fields. – Amer. Assoc. Petrol. Geolog., Tulsa 1970.

Halbouty, M. T. & King, R. E. & Klemme, H. D. et al.: Factors affecting formation of giant oil and gasfields, and basin classification. - In: Geology of giant oil accumulations. - Bull. Amer. Assoc. Petrol. Geolog., Tulsa 1972.

Haun, J. D. & Le Roy, L. W.: Subsurface geology in petroleum exploration. - Colorado School of Mines, 1958.

Haun, J. D.: Origin of petroleum. - AAPG reprint series No. **1**, Golden 1971.

Hedberg, H. D.: Geologic aspects of origin of petroleum. - Bull. Amer. Assoc. Petrol. Geolog., **48**/11, 1755, Tulsa 1964.

Hedberg, H. D.: Continental margins from viewpoint of the petroleum geologist. - Bull. Amer. Assoc. Petrol. Geolog., **54**/1, 3, Tulsa 1970.

Hedemann, H. A.: Geologische Auswertung von Temperaturdaten aus Tiefbohrungen. - Erdöl und Kohle, **20**/5, 337, 1970.

Hendricks, T. A.: Resources of oil, gas and natural liquids in the United states and the World. - U.S. Geol. Surv. Cire. **522**, 1965.

Heroy, W. B.: Unconventional methods in exploration for petroleum and natural gas. - SMU Symposium **87**, Dallas 1969.

Hintze, W. E.: Depiction of faults on stratigraphic isopach maps. - Bull. Amer. Assoc. Petrol. Geolog., **55**/6, 871, Tulsa 1971.

Hitchon, B. & Horn, M. K.: Petroleum indicators in formation waters from Alberta, Canada. - Bull. Amer. Assoc. Petrol. Geolog., **58**/3, 464, Tulsa 1974.

Hobson, G. D.: Some fundamentals of petroleum geology. - Oxford University Press, London 1954.

Hobson, G. D.: Faulting and oil accumulation. - Journ. Inst. Petrol., **42**/385, 23, 1956.

Hobson, G. D.: Factors affecting oil and gas accumulations. - Journ. Inst. Petrol., **48**/165, 1962.

Hobson, G. D.: Petroleum accumulation. - Journ. Inst. Petrol. **59**/567, 139, 1973.

Hobson, G. D. & Tiratsoo, E. N.: Introduction to petroleum geology. - Scientific Press, Beaconsfield 1975.

Hollenshead, C. T. & Pritchard, R. L.: Geometry of producing Mesaverde sandstones, San Juan Basin. - In: Geometry of sandstone bodies. - Amer. Assoc. Petrol. Geolog., Tulsa 1961.

Hopping, C. A.: Palynology and the oil industry. - Rev. Palaeobotan. Palynol., **2**/23, 1967.

Horowitz, A. S. & Potter, P. E.: Introductory petrography of fossils. - Springer, Berlin 1971.

Howell, J. V.: Structure of typical American oil fields. - Amer. Assoc. Petrol. Geolog., Tulsa 1948.

Hoyer, W. G. & Rumble, R. C.: Field experience in measuring oil content, lithology and porosity with a high-energy neutron-induced spectral logging system. - Journ. Petrol. Technolog., **17**/7, 801, 1965.

Hoyt, J. H.: Chenier versus barrier, genetic and stratigraphic distiction. - Bull. Amer. Assoc. Petrol. Geolog., **53**/2, 299, Tulsa 1969.

Hubbert, M. K.: History of petroleum geology and its bearing upon present and future exploration. - Bull. Amer. Assoc. Petrol. Geolog., **53**/2, 299, Tulsa 1969.

Hubbert, K. M.: Application of hydrocynamics to oil exploration. - Proc. 7th World Petrol. Congr. Mexico, **1**, 59, 1967.

Hughes, R. V.: Oil property valuation. - John Wiley & Sons, New York 1967.

Hull, C. E. & Warman, H. R.: Asmari oilfields of Iran. - In: Geology of giant petroleum fields. - Amer. Assoc. Petrol. Geolog., Tulsa 1970.

Irving, E. & North, F. K. & Couillard, R.: Oil, climate and tectonics. - Canad. Journ. Earth Sci., **11**/1., 1, 1974.

Jakosky, J. J.: Exploration geophysics. - 2nd Ed. Los Angeles 1950.

Jones, M. P. & Fleming, M. G.: Identification of mineral grains. - Elsevier, 1965.

Kamen-Kaye, M.: Geology and productivity of Persian Gulf synclinorium. - Bull. Amer. Assoc. Petrol. Geolog., **54**/12, 2371, Tulsa 1970.

Kartsev, A. et al.: Geochemical methods of prospecting and exploration for petroleum and natural gas. - Univ. California Press, 1959.

Keelan, D. K.: A critical review of core analysis techniques. - Journ. Canad. Petrol. Technol., **11**/2, 42, 1972.

Keller, G. H. & Frischknecht, F. C.: Electrical methods in geophysical prospecting. - Pergamon Press, London 1966.

Keller, G. V.: Electrical prospecting for Oil. - Quarterly Colorado School of Mines, 1968.

Kidwell, A. L. & Hunt, J. M.: Migration of oil in recent sediments of Pedernales, Venezuela. - In: Habitat of oil. - Amer. Assoc. Petrol. Geolog., Tulsa 1958.

Kokesh, F. P. et al.: New approach to sonic logging and other acoustic measurements. - Journ. Petrol. Techn., **17**/3, 283, 1965.

Kramer, K. H.: Erdöl-Lexikon. - 5. Ed. Dr. Alfred Hüthig Verlag, Heidelberg 1972.

Krejci-Graf, K.: Erdöl. - Springer Verlag, 1955.

Kroepelin, H.: Geochemical prospecting. - RP **3**, 7th World Petrol. Congr., Mecixo 1967.

Krumbein, W. C. & Sloss, L. L.: Stratigraphy and sedimentation. - Freeman, San Francisco 1963.

Kumar, J.: New chart offers fast permeability estimate. - World Oil, **38**,

Feb. 1971.
Kvenvolden, K. A.: Molecular distributions of normal fatty acids and paraffins in some Lower Cretaceous sediments. - Nature, **209**, 573, 1966.
Laurien, H.: Taschenbuch Erdgas. - Oldenbourg Verlag, München/Wien 1966.
Levorsen, A. J.: Stratigraphic type oil fields. - Amer. Assoc. Petrol. Geolog., Tulsa 1941.
Lindholm, R. C.: Detrital dolomite in Onondaga Limestone (Middle Devonian); its implication to the dolomite question. - AAPG Reprint series No. **4**, Tulsa 1972.
Lindsey, J. P.: Bright spot; a progress report and look ahead. - World Oil, **81**, April 1974.
Link, W. K.: Significance of oil and gas seeps in world oil exploration. - Bull. Amer. Assoc. Petrol. Geolog., **36**/8, 1505, Tulsa 1952.
Lintz, J.: Remote sensing for petroleum. - Bull. Amer. Assoc. Petrol. Geolog., **56**/3, 542, Tulsa 1972.
Lovejoy, W. F. & Homann, P. T.: Methods of estimating reserves of crude oil, natural gas and natural gas liquids. - John Hopkins Press, Baltimore 1965.
Lowry, W. D.: North American geosynclines − test of continental drift theory. - Bull. Amer. Assoc. Petrol. Geol., **58**/4, 575, Tulsa 1974.
Lynch, E. J.: Formation evaluation. - Harper and Row, New York 1962.

Magub, T. A.: The Geoseis system. - APEA Journ. **12**/1, 28, 1972.
Martin, R.: Papaeogeomorphology and its application to exploration for oil and gas. - Bull. Amer. Assoc. Petrol. Geolog., **50**/10, 2277, Tulsa 1966.
Matthews, C. S. & Russell, D. G.: Pressure build up and flow tests in wells. - Soc. Petrol. Eng. AIME, New York 1967.
Mayer-Gürr, A.: Grundfragen der Erdölförderung. - Herrnhausen Verlag, Hannover 1944.
Mayer-Gürr, A.: Erschließung und Ausbeutung von Erdöl- und Erdgasfeldern. - In: Lehrbuch der angewandten Geologie. - (Ed. Bentz, A. & Martini, H. I.) **II**/1, Ferd. Enke Verlag, Stuttgart 1968.
Meinhold, R.: Geophysikalische Meßverfahren in Bohrungen. - Akadem. Verlagsges., Leipzig 1965.
McKenzie, D. P.: Plate tectonics of the Mediterranian region. - Nature, **226**, 239, 1970.
Megill, R. E.: An introduction to exploration economics. - Petroleum Publishing Co., Tulsa 1971.
Milner, H. B.: Sedimentary petrography. - Allen and Unwin, 1962.
Moody, J. D.: Petroleum demands of future decades. - Bull. Amer. Assoc. Petrol. Geolog., **54**/12, 2239, Tulsa 1970.
Moore, G. T.: Interaction of rivers and oceans. - Bull. Amer. Assoc. Petrol. Geolog., **53**/12, 2421, Tulsa 1969.
Murray, G. E.: Salt structures of Gulf of Mexico basin-a review. - In: Diapirism and diapirs. - Amer. Assoc. Petrol. Geolog., Tulsa 1968.

Nettleton, L. L.: Gravity and magnetics for geologists and seismologists. - Bull. Amer. Assoc. Petrol. Geolog., **46**/10, 1815, Tulsa 1962.
North, F. K.: Characteristics of oil provinces. - Bull. Canad. Petrol. Geolog., **19**/3, 601, 1971.

Pennebaker, A.: Vertical net sandstone determination for isopach mapping of hydrocarbon reservoirs. - Bull. Amer. Petrol. Geolog., **56**/8, 1520, Tulsa 1972.

Perrodon, A.: Géologie du pétrole. - Presse Université de France, Paris 1966.

Perry, E. A. & Hower, J.: Late stage dehydration of deeply buried pelitic sediments. - Bull. Amer. Assoc. Petrol. Geolog., **56**/10, 2013, Tulsa 1972.

Philippi, G. T.: On the depth, time and mechanism of petroleum generation. - Geochim. and Cosmochim. Acta, **29**/9, 1021, 1965.

Pirson, S. J.: Handbook of well-log analysis. - Prentice-Hall, Englewood Cliffs, 1963.

Pixler, B.: Formation evaluation by analysis of hydrocarbon ratios. - Journ. Petrol. Technol., **21**/6, 665, 1969.

Porfirov, V. B.: Inorganic origin of petroleum. - Bull. Amer. Assoc. Petrol. Geolog., **58**/1, 3, Tulsa 1974.

Powers, M. C.: Fluid-release mechanisms in compacting marine mud-rocks and their importance in oil exploration. - Bull. Amer. Assoc. Petrol. Geolog., **51**/7, 1240, Tulsa 1967.

Pugh, W. E. & Preston, B. G.: Bibliography of stratigraphic traps. - Seismogr. Service Corpor., Tulsa 1953.

Rade, J.: Otway basin, Australia; use of calcareous nannoplancton and palynology to determine depositional envirenment. - Bull. Assoc. Petrol. Geolog., **54**/11, 2196, Tulsa 1970.

Rees, F. B.: Methods of mapping and illustrating stratigraphic traps. - AAPG Memoir **16**, Amer. Assoc. Petrol. Geolog., Tulsa 1972.

Rittenhouse, G.: Stratigraphic-trap classification. - Memoir **16**, Amer. Assoc. Petrol. Geolog., Tulsa 1972.

Riboud, J. & Schuster, N. A.: Well-logging techniques. - R. P. **7**, 8th World Petrol. Congr., Moscow 1971.

Robert, M.: Géologie du pétrole. - Gauthier-Villars, Paris 1959.

Royal Dutch/Shell: Standard Legend. - N. V. de Bataafsche Petroleum Matschappij, The Hague, 1958.

Rudman, A. J. & Lankstone, R. W.: Stratigraphic correlation of well – logs by computer techniques. - Bull. Amer. Assoc. Petrol. Geolog., **57**/3, 577, Tulsa 1973.

Rühl, W.: Entölung von Erdöllagerstätten durch Sekundärverfahren. - Beih. Geolog. Jahrb. **4**, Hannover 1952.

Russell, W. L.: Interpretation of neutron well logs. - Bull. Amer. Assoc. Petr. Geolog., **36**/2, 213, Tulsa 1952.

Ryan, J. T.: An analysis of crude-oil discovery rate in Alberta. - Bull. Canad. Petrol. Geolog., **21**/2, 219, 1973.

Schatzow, N. J.: Bohren auf Erdöl und Erdgas. - VEB Verlag Leipzig 1965.

Schönwälder, G.: Erdöl in der Geschichte. - Hüthig & Dreyer Verlag, Mainz und Heidelberg 1958.

Schwarzacher, W.: Sedimentation in subsiding basins. - Nature, **241**, 1349, 1966.

Shelton, J. W.: Role of contemporaneous faulting during basin subsidence. - Bull. Amer. Assoc. Petrol. Geolog., **52**/3, 399, Tulsa 1960.

Shepard, F. P.: Submarine geology. - Harper & Row, New York 1973.

Shimoyama, A. & Johns, W. D.: Formation of alcanes from fatty acids in the presence of $CaCO_3$. - Geochim. et Cosmochim. Acta, **36**/1, 87, 1972.

Silverman, S. R.: Influence of petroleum origin and transformation on its distribution and redistribution in sedimentary rocks. - PD **1**, 8th World Petrol. Congr., Moscow 1971.

Simpson, J. P.: What's new in mud engineering. - World Oil, 135 (April)

and 118 (May) 1967.
Sitter, L. U. de: Diagenesis of oilfield brines. - Bull. Amer. Assoc. Petrol. Geolog., **31**/11, 2030, Tulsa 1947.
Sitter, L. U. de: Structural geology. - McGraw-Hill, 1964.
Smith, D. A.: Theoretical considerations of sealing and non-sealing faults. - Bull. Amer. Assoc. Petrol. Geolog., **50**/2, 363, Tulsa 1966.
Sullwold, H. H.: Turbidites in oil exploration. - In: Geometry of sandstone bodies. - Amer. Assoc. Petrol. Geol., Tulsa 1961.
Stahl, W.: Kohlenstoff-Isotopenanalysen zur Klärung der Herkunft nord-westdeutscher Erdgase. - Diss. Clausthal 1968.
Stahl, W.: Carbon isotope fractionations in natural gas. - Nature, **251**, 134, 1964.
Staplin, F. L.: Sedimentary organic matter, organic metamorphism and oil and gas occurence. - Bull. Canad. Petrol. Technolog., **17**/1, 47, 1969.
Stevens, W. F. & Thodos, G.: New method for estimating primary oil reserves. - World Oil, 113, Dec. 1961.
Stout, J. L.: Pore geometry as related to carbonate stratigraphic traps. - Bull. Amer. Assoc. Petrol. Geolog., **48**/3, 329, Tulsa 1964.

Tapper, W.: Caliper and temperature logging. - In: Subsurface methods in petroleum geology. - Color. School of Mines, **439**, 1951.
Tarrant, L. H.: Geophysical methods used in prospecting for oil. - Applied Science Publ., 1973.
Tickell, F. G.: The techniques of sedimentary mineralogy. - Elsevier 1965.
Timko, D. J.: Recent trands in formation evaluation. - World Oil, 97, June 1968.
Tiratsoo, E. N.: Natural Gas. - Scientific Press, Beaconsfield 1972.
Tiratsoo, E. N.: Oilfields of the world. - Scientific Press, Beaconsfield 1973.
Tissot, B. & Califet-Debyser, Y. & Deroo, G. & Oudin, J. L.: Origin and evolution of hydrocarbons in early Toarcien shales. - Bull. Amer. Assoc. Petrol. Geolog., **55**/12, 2177, Tulsa 1971.
Tittle, C. W. & Allen, L. S.: Theory of neutron logging. - Geogphysics, **31**, 314, 1966.
Tixier, M. P.: Modern log analysis. - Journ. Petrol. Technol. **14**/1327, 1962.
Trask, P. D.: Origin and environment of source sediments of petroleum. - Gulf Publish. Co., Houston 1932.
Trask, P. D. & Patnode, H. W.: Source beds of petroleum. - Amer. Assoc. Petrol. Geolog., Tulsa 1942.

Walters, R. F.: Buried Pre-Cambrian hills in northeastern Barton County, central Kansas. - Bull. Amer. Assoc. Petrol. Geolog., **30**/5, 660, Tulsa 1946.
Weeks, L. G.: Habitat of oil. - Amer. Assoc. Petrol. Geolog., Tulsa 1958.
Weeks, L. G.: Marine geology and petroleum resources. - PD **2**/3, 8th World Petrol. Congr., Moscow 1971.
Wiebe, W. A. V.: Oil fields in the United States. - McGraw-Hill, New York 1930.
Williamson, H. F. & Daun, A. R.: The American petroleum industry; the age of illumination. - Northwestern Univ. Press, Evanston 1959.
Williamson, H. F. & Andreano, R. L. & Daum, A. R.: The American Petroleum industry; the age of energy. - Northwestern Univ. Press, Evanston 1963.
Wing, R. S. & MacDonald, H. C.: Radar geology-petroleum exploration technique, Eastern Panama and Northwestern Columbia. - Bull. Amer. Assoc. Petrol. Geolog., **57**/5, 825, Tulsa 1973.

Wyllie, M. R. J.: The fundamentals of electric log interpretation. - Academic Press, New York 1963.

Young, A. & Galley, J. E.: Fluids in subsurface environments. - Amer. Assoc. Petrol. Geolog., Tulsa 1965.

Zemanek, E. & Glenn, E. & Norton, L. J. & Caldwell, R. L.: Formation evaluation by inspection with the borehole televiewer. - Geophysics, **35**/2, 254, 1970.

Index

absorber 109
absorption 18
accumulation of oil 32, 45, 59
acetate copy 120
acetic acid 120
acid frac 92
acoustic log 145, 146, 150, 152, 155
acoustic properties 65
acoustic velocity 66
aestuarine formation 151
airborne magnetics 60, 61, 62
airfotogrammetry 61
airpollution 39
alcanes 22, 23
alchemists 4
algae 138
americium/beryllium 107, 150
amino acids 16, 17
anhydrite 38, 52, 114
analyse of cuttings 112
anticlinal structure 39, 40, 51, 52, 53, 54, 61
anticlinal theory 13
antiseptic 7, 8
API – grade 37
apparent depth 118
apparent thickness 31, 134
ape board 73, 82, 84
aquifer storage 154
arenite 138
argon 36
areometer 36, 92
ark of Noah 2
aroclor 121
aromates 22, 24, 37
Asmari limestone 139
asphaltic crude 25
asphaltite 20
asphalt powder 91
asphalt prop 49
attenuation of sound 150
atmospheric pressure 19
attapulgite 87, 91, 94
average grain size 33

backfall 116, 119
back scattered gamma rays 107, 149
balcony field 42

banner of Ur 1
baroid press 93
baryte 89, 92
bees wax 119
below disconformity 53
benzene 24
beta counter 94
bioherm field 39
bit 81, 82
bituminous limestone 20
bituminous slate 20
blasting 65
block 70, 72
blow out 33, 73, 74, 88, 89, 94, 97, 106, 156
boghead coal 21
bore hole logging 29, 56, 58, 64, 69, 71, 115, 123, 125, 131, 135, 136, 140, 141, 151, 152, 153
borehole wall 161
bottom anchor 157
bottom choke 162
bottom water 32, 33
Bouguer anomaly 63
brakish formation 151
brass ring method 93
bromide 18, 34
breccia 52
bulk porosity 30, 134
bulk thickness 31
bullet perforation 162
buried hill 39, 53, 54, 63
burning rock 10
bushings 79, 80
butane 22, 35
button bit 83

C 12/C 13 ratio 15, 21
cable test 71, 156, 157
calcite 92, 114, 137, 138
caliper log 151, 153
capillary sieve 104
cap rock 23, 52
cap rock field 51
capsule 164
carbide 15
carbonaceous cementation 29
carbobitumina 20
carbon dioxide 35, 97

carbon disulphide 20
carbonic acid 97
cased hole 146, 148
casing 69, 73, 164
casing head 75, 92
cat head 75, 78, 80, 85
St. Catherine oil 8
caustobiolithe 62
caved out formation 151
cavern 18, 25
cavities 18
cementation 29, 96, 138, 150, 163
cementing shoe 162
cement slurry 163
centralizer 162, 163
chemical fracturing 26, 30
chloroform test 111
choke 164
christmastree 69, 165
chromatograph 104, 109
claystone 113
closed in pressure 156, 159, 161
closure 33, 45
CMC (carboxylmethylcellulose) 91
coal 20
cobalt-60 107
coccolithic algae 121
cold wire detector 97, 99, 108
collimator 140, 149
colloidal structure 94
colloidity 91
column 104
cooking point 34
copper 17
commercial value 34
compaction 18, 19
compass 151, 153
completion 56, 69, 155, 156, 165
completion log 151
complex hydrocarbons 22
condensate 20, 23, 35, 103, 135
condensate field 103
conductive mud 146
conductivity 94, 145
conductivity bridge 93, 94, 97
conglomerate 63, 152
connate water 19, 33, 34
conodonts 118
conservation 18
consolidated sand 94
contamination 91
continous dipmeter 134, 153
control panel 74, 80
conventional wiggle 65

coral limestone 137
coral reef 138
core analysis 123, 131
core barrel 73, 123, 125, 126, 129,
    130, 131
core catcher 126, 129, 131
core container 131
core description 131
core drilling 123, 125
core loss 123
core tester 132, 133
coring a pay 124
cracks 135, 138, 150
crevices 150
cross potential 97, 105
crown 70, 73, 79, 85
crown block 83
crown load 85
crude oil 20, 36 ff.
crypton 36
cuttings 71, 78, 83, 84, 90, 91, 93,
    97, 99, 109, 110, 111, 118, 120
cycloalcanes 22, 24
cyclohexan 24
cyclone 93
cyclopentan 23, 24

Darcy 26, 136
Darcy formula 136
dark field condenser 120
dead porosity 29, 135
dead organic material 18
decane 22
decomposition 18
degasser 90, 92
dehydration 18
density 90, 91, 92, 108
density log 107, 153, 155
density of mud 106, 107
density recorder 106
department for exploitation 55
department for exploration 55
department for technical services 55
deriviation log 151
derrick 68, 73, 79, 85, 86, 87
desander 89, 93
desulfovibrio 23
detector 97, 104, 109
development of a field 32
deviation of a bore hole 123
diagenesis 48, 137, 138
diamond 22
diamond bit 83
diamond crown 123, 126, 128

differential load 84
dipmeter 153
dip of a well 31
direct coring 124
directional hole 68
disconformity 47, 52, 53, 59, 65, 152, 153
disconformity field 46
disintegration 118
dissolved gas 62, 96
disulphide 24
doghouse 75, 77, 83, 106
dogleg 141
dolomite 112, 138
downthrust fault 43
downturned block 45
Drake 5
drawworks 72, 73, 79, 80
drill collars 73, 82
driller 55
driller's stand 74, 80
drilling branch 55
drilling mud 58, 90, 144, 162
drilling fluid 90
drilling crew 70
drilling string 72, 77, 78, 79, 82, 87, 125, 129
drilling technique 56, 70
drilling tool 82
drilling rate 117
drilling rig 64, 72
drillometer 82, 83, 84
drillpipe 79, 81, 127, 162
drill recorder 81, 117
drillsite geologist 58
drill stems 73
drill stem test 71, 156, 157, 159
dry gas 35
dynamite 65

economy 3
edge water 30, 32, 33
effective porosity 30
effective thickness 31, 134, 153
electric log 152
electric pressure sensor 101
electric resistivity 144
electronic circuit 140
elevator 70, 72, 73, 79, 81, 84
emitter coil 147
energy 3
energy level of a well 164
engine 69, 70, 72, 73
essential oils 7

eruption cross 165
ethane 22, 35
evaluation of pay formation 155
evaporation of crude 49
exhaust gas 101, 109
exhausted well 12
exchange of oxygen 17
exploration 55, 58, 59
exploitation 55, 56

facies trap 98
facies type field 48, 52
fault 61, 66, 152
fault plane 42, 45
fault type field 39, 42, 51
fatty acid 16, 17
faunule 119
field above a disconformity 51, 54
field below a disconformity 50, 54
final casing 156, 164
financial department 55
fire worshippers 9
first production test 36
fishing job 141
fissures 25, 26, 92, 135, 136, 138, 150
fixative spray 122
fixing solution 142
flame ionisation detector 99, 104
flame point 36, 38
flame thrower 6
flaring gas off 36
flow channel 157
float for mud gas 97
flow pressure 156
flowing process 30
flow test 159
flow valve 156
fluid loss 91, 92
flushed zone 145
flushing 131
focussed resistivity methods 134, 147, 152
food 11
foraminifera 118, 120
formation interval tester 161
formation of reservoirs 59
formation pressure 92, 94, 156
formation water 32, 34, 144, 155
formic acid 120, 121
fossil resins 21
fossil wax 21
fotogeology 61
fractured formation 25, 26, 96

fresh water mud 145, 146, 148
frequency curve 29
fundamental laws of petroleum
    exploration 59

gas 135
gas alarm 108
gas anchor 95, 124, 164
gas bearing pay 132
gas bubbles 109, 112
gas cap 32, 40
gas chromatograph 99
gas container 132, 133
gas detector 56, 62, 98, 132, 133
gas field 33
gas float 99
gas lab 78
gas logging 62, 84, 101, 109, 123
gas/oil ratio (GOR) 33
gasoline 37, 38
gas peak 117
gas sniffing 62
gas stirrer 107
gas tail 108, 109
gas trace 109
gas volcanoe 12
galena 92
galvanometer 142
gamma activity 94
gamma counter 94
gamma- gamma logging 107
gamma radiation 149
gamma ray log 107, 146
gamma ray diagram 154, 161
gamma ray neutron log 148
gamma ray of capture 150
gamma source 107
gear box 88
gel strength 91, 108
genesis of petroleum 58
genesis of sulphur 23
genuine sea water 18
geochemist 60
geoelectrics 64, 66
geological lab 91
geomagnetics 60
geophone 65, 66
geosyncline 39, 45, 50
geophysics 55, 60
ginpole 73
giant field 42
glacial deposit 61
good bond 150, 163
goose neck 72, 75, 78, 79, 87

GOR (gas/oil ratio) 34
graben structures 39, 44, 47
grade of roundness 137
grain model 26, 27
grain size analysis 29, 134
granite wash 53
granular pay 27
graphical plot 32, 134
graphite 22
gravimetry 53, 60, 63, 66
gravity survey 36, 55
greek fire 2, 5, 6
ground water 138
gypsum 38, 52, 91, 114

handy oil 38
hampelmann 75
hard formation 120, 127, 129, 130
hard streak 31, 127, 154
heater treater 38
heavy hydrocarbon gas 33, 62
helium gas 36, 104
hematite 92
heptane 21
hemin 16
hexane 22, 35
high frequency log 147
high gravity crude 38
hook 72, 79
hook wall anchor 156, 157
horse head 164
hot water method 110, 112
hot wire detector 97, 109
hybrid hydrocarbons 22
hydraulic fracturing 26
hydraulic line 73
hydraulic pressure 161
Hydril preventer 75, 76, 77
hydrocarbon gas 97, 100, 103
hydrocarbonic acid 35
hydrochloric acid 114, 121, 132
hydrogen gas 97, 108
hydrogen peroxide 118
hydrogen sulphide 24, 35, 97, 101
hydrostatic pressure 91

illumination 4
immersion lens 122
imperfect bond 150
impervious boundary 32
induction electric log 148
induction log 145, 147, 148
infra red light 112
infra red detector 101, 109

initial close in pressure 159
inlet valve 88
inner surface 33
inorganic origin 15
ion exchange 94
insoluble residue 121
interior surface 29
intercalation 30
intermediate crude 37
interpretation 65
interstitial water 18
inverse curve 152
iron oxide 17
iron silicate 137
jet bit 73
jet nozzles 83, 88
jodine 18, 34

kelly 72, 73, 75, 78, 79, 81, 84, 99
kerobitumina 20
kerosene 3, 4
kerosene lamp 5
key hole 141

laboratory 57
laboratory analysis 56
laboratory test 136
lattice type rig 85
lateral curve 152, 154
lead acetate 35, 101
leasing man 68
legal department 55
limestone 52, 112, 138, 146
limy marl 113
limnic formation 151
LNG (liquid natural gas) 9
liner 156, 162
line theory 13
LPG (liquid petrol gas) 5
lipoid 20
load 130
logging cable 140, 157
logging cabin 142
logging camera 141, 142
logging car 140, 141
logging engineer 142
logging program 142, 157
logging specialist 57
logging speed 142
long normal 154
Lorac 61
lorry 64
lost circulation 94
low gravity crude 38

lubricating oil 38, 101
luminescence 110, 111, 132
Lussagnet 40

magnetic anomaly 61, 62
magnetic field 61
magnetic map 66
magnetic tape 65
manifold 73
marine basin 151
marine facies 152
marketing 56
marl 127
Marsh funnel 91
mast 72, 75
measuring bridge 93
measuring cell 99
medium gravity crude 38
medium diameter 135
mercaptane 24, 39
mercury pump 136
mesofossil 119
metamorphic rock 139
metamorphism 59
metamorphosis 137
methane 21, 22, 35, 36, 97, 103
methylcyclohexane 23
micas 92
microbe 18
microfossil 118, 119
microlaterolog 134, 146, 147, 155
micropaleontology 56, 57, 118, 125
migration 18, 19, 32, 59
millidarcy 26, 136
L. Mintrop 63
mixer 109
mixing tank 70, 73
moisture logging 107
molecular sieve 105
monazite sand 149
monochloracetic acid 120, 121
monocolor detector 99
mono-duplex-detector 99
montmorillonite 87, 91, 94
mouse hole 75, 81, 84
mud additives 91, 92
mud balance 92
mud cake 151
mud circulation 87, 141, 144
mud column 156, 159
mud gas analysis 97, 105
mud gas detector 96, 106
mud gas logging 71, 78, 88, 90,
    105, 106

mud gas recorder 84
mud gas stream 101
mud filtrate 93, 144, 145, 156, 159
mud funnel 75, 77, 87, 88, 164
mud hose 87
mud level recorder 92, 106, 107
mud pit 91, 93
mud pressure 82, 84
mud pump 70, 72, 73, 87
mud resistivity 145
mud return ditch 70, 75, 77, 87, 99
mudstone 113, 127
mud tanks 73
mud valve 161
multiple pay 30
mumia 1, 7, 8

nannofossil 121
napalm 6
naphta 1
naphthenic aromates 22
naphthenic crude 37
naphthenes 22, 37
naphthene compounds 38
nappes 46
natural gamma ray log 149
natural gas 20, 35
natural hydrocarbons 20
natural potential 147
neutron log 107, 146, 148, 154, 155
neutron detector 108
neutron/gamma log 107, 146
neutron/neutron log 107, 149
nitric salts 16
nitrogen compounds 25
nitrogen 35, 36, 97
normal fault 50, 52
non compressible fluid 136
non conductive material 147
non conductive mud 148
non consolidated sandstone 127
nuclear logging 145, 146, 155

offset well 116
oil and gas analysis 58
oil and gas detection 90
oil base mud 145, 146, 148
oil bleeding core 59
oil impregnation 132
oil prospection 59
oil pump 164
oil reserves 12
oil seep 59
oil trace 110

oil transportation 60
oil trap 19
oil wet pay 19, 25
open hole test 36, 156
optical measuring bridge 102
organic origin 15
origin of hydrocarbons 15
origin of porosity 137
ostracods 118
outlet valve 88
overburden pressure 19
overflow 87, 88, 90, 97
overlap field 46, 47, 53
overpressure 92
overthrust fault 46, 53
overthrust field 52
oxidation of crude 49
oxidation zone 18
ozokerite 20

packer 161
packer shoe 159
paraffin 22, 23, 37
paraffin oil 5
paraffinic crude 37
parsees 9
path of migration 19
pay 25, 55
pay density 29
pay detection 97
pay evaluation 140, 153, 155
pay formation 42, 69, 133, 137, 148
peak correlation 152
peat 20
pellistor 100, 101
pendulum 151, 153
penetration rate 84, 91, 117
pentane 22, 35
pentane/methane ratio 103
perforated liner 156
perforation 30, 69, 162
perforator 162
permeability 25, 26, 30, 32, 133,
    134, 135, 136, 137, 138, 159
permeability analysis 57
permeable layer 32
permeameter 57
petrobitumina 16, 20
petrography 119
petroleum hydrocarbons 20, 59
petrographic analysis 112, 113
petrographic details 147
petroleum 1
petrolog 115, 117

petrophysical properties 142
petrophysics 57
pinch out 53, 61
pipe rack 71, 73, 84
piston 88, 161
planctonic organisms 16
platform 69, 75, 78, 82
platinum filament 97, 101
plug 163
pocket lens 132
pollution 6, 90
polymerisation of crude 49
polymorphic structures 50
polysulphides 1, 7
polysulphuric compounds 39
poor bond 163
pore contents 18, 131, 148, 153
pore space 26, 27, 33, 34
porosity 25, 26, 29, 32, 133, 134,
    135, 146, 153
porosity analysis 57
porous formation 151
porous zone 147
porphyrines 16, 17
potassium 90, 91, 149
pour point 36, 38
power slips 79
precipitation 138
pressure build up 159
pressure chart 156
pressure diagram 157
pressure drop 33, 34
pressure gradient 19
pressure recorder 156, 159
pressure sensor 73, 156
preventer 83, 95
primary porosity 138, 139
primar pore filling 33
production control 32
production engineer 71, 136
production engineering 55, 56
production packer 162, 164
production technique 57
propane 22, 35, 103
prospective pay 67
proteins 11, 20
pulsation dampener 88

quartz 114
quartzite 114, 146
quebracho 91
St. Quirinus oil 8

radioactive source 140

radioactivity 149
rake 82
ram 74
rate of penetration 81, 84, 119, 124
rate of settling 42
rat hole 75
reactivation of formation pressure 34
real depth 118
reamer 73, 82
reaming a bore hole 130
receiver 150
receiver coil 147
reconstruction of fossils 120
recorder 73
reduction zone 18
reef 54
reef limestone 139
reflecting horizon 65
reflector 65
reflexion seismics 63, 64, 66
refraction diagram 64
refraction seismics 63, 66
refraction index 120
refining process 38
regional trend 64
registration car 64
religion 9
resin 25
residual oil 124
resistivity 66, 90
resistivity logging 145, 152, 155
resistivity of rock 146, 154
resolution 147, 152
return ditch 73
reverse fault 42, 43, 45
reversing 105
rig 68
ripening process 18
rise time 117
road conditions 141
rock salt 66, 90
roller bit 82
rope 73, 79, 83
rotary bit 73, 83
rotary derrick 72
rotary hook 73, 79
rotary hose 72, 73, 75, 87
rotary line 73, 80
rotary table 72, 73, 77, 78, 80, 84,
    85, 131
rotating speed 130
roughneck 55, 77, 78, 162
round trip 85, 108
rotary tongue 78, 80, 85

royalties 2, 4
rubber packer 156
rubber plug 162
rubber seal 159
runaround 73, 85
rupture disc 162, 163

saline water 146
salinity 144
saltbearing formation 94
salt contents 94
salt diapir 51
salt dome 45, 50, 51, 62, 64, 66, 92
salt dome fields 39
salt mud 145, 146
sample 70, 117
sampler 55, 56, 58, 91
sand contents 93, 114, 116
sand grain 115
sand line 31
sandstone 114, 146
sandwich formation 31
saturation 33
sawdust 92
Schaffer preventer 74
scintillation counter 107, 149
scintillometer 107, 149
scintillation crystal 149
scraper 163
screen 119
secondary porosity 138, 139
secondary production 32
sedimentary basin 32, 62
sedimentation 137
sedimentologist 57
sedimentology 56, 58
segregation 32
seepage 12, 59
seismic recorder 64
seismic investigation 53
seismics 60
self potential 144, 148, 154, 157, 161
settling point 73
settling tank 70, 87
shale 113
shale shaker 70, 75, 78, 87, 88, 90, 99, 108, 117, 144
shaling out 50, 51, 54
shaped sharge 162
shelf 15
shooting cores 129
short normal 134, 144, 152, 154
shot hole 64, 65

shrinkage 53
shut in pressure 159
side wall coring 127, 130, 131
silica 139, 146
siliceous cementation 29
single test 156, 157
slips 77, 79
slip rings 79
slush pumps 73, 83, 108
Sodom and Gomorrha 10
soft formation 129, 130, 151
solid contents 93
solidification 138
solid sulphur 24
solvent 111
sonic log 148, 153, 155
sorting of grain sizes 137
sour gas 35
spersene 91
spontaneous potential 147
spring valve 159
squeezing 19
stand pipe 73, 164
steel bullets 162
stimulation of a well 26
stirrer 88
straddle test 156, 157
stratigraphic analysis 116
stratigraphic comparison 140, 151, 152
stratigraphic type field 39, 48
streaming potential 34
stripping 39
structural drilling 67
stuffing box 78
substructure 72, 75
subsurface water 138
surface activity 18, 25
surface gas logging 62
suction line 88, 97
suction tank 70, 87
sulphides 24
sulphide detector 101
sulphur 36, 37
sulphur contents 38
sulphur supply 25
sulphuric compounds 23
sulphuric gases 35
surface active transistor 100
surface tension 19, 34
sweet gas 35
swivel 70, 72, 73, 75, 78, 79, 83, 87

tar 37

tar pits 49
tail pipe 156
team work 55, 71
tectonic analysis 140
tectonics 138
telluric survey 64
telescopic mast 86
terrace field 92
terrestrial formation 151
tertiary production 32
testing 56
tetralin 22
thermal conductivity 100
thermistor 99
thickness 133
thickness of pay 30
thin section 120
thiophene 24
three cone bit 83
Thyrsus oil 8
thorium minerals 149
tiltmeter 151
toluene 24, 121
tongue 78, 80
tool joint 75
topography 55, 60, 63
total weight 81, 82, 84
tower of Babel 1
trailer mounted rig 6, 8
transgression 46, 53
transformation 18
transition zone 33, 34
trap 19, 22
travelling block 73, 75, 79, 85
true porosity 135
true resistivity 145
true thickness 31, 134
tubing 64, 164
tungsten carbide 128
tugboat 69
turbine drilling 83

unconsolidated sediment 19
underground storage 7, 35

underground water 18
ultrasonic shaker 121
ultraviolet light 110, 132
upturned block 45
uranium salt 149
Uras gas detector 98

vacuum pump 132, 133
valves 73
vanadium 17
variable area 65
variable density 65
velocity of sound 150
viscosimeter 91
viscosity of crude 25
viscosity 90, 91
viscosity of mud 89
volatile contents 131
vuggy formation 96
vugular limestone 25

water loss 90, 96
water wet 19, 25
wax 37
way of migration 32
weighted mud 92
weight dropping 65
weight indicator 73
weight on the bit 81, 82
well completion 155, 162
well head 69, 162, 164
well site geologist 55, 57, 162
well site lab 131
well stimulation 26
wet gas 35
Wheatstone bridge 97
wildcat 56, 57, 60, 68, 116, 155
winch 69, 81
wire line coring 128
working platform 72

yeast 11

zone of infiltration 145